Reworking the Workshop

Workshop

Math and Science Reform in the Primary Grades

DANIEL HEUSER

HEINEMANN
Portsmouth, NH

Heinemann
A division of Reed Elsevier Inc.
361 Hanover Street
Portsmouth, NH 03801–3912
www.heinemann.com

Offices and agents throughout the world

The author and publisher wish to thank those who have generously given permission to reprint borrowed material:

Figure 9–12 is reprinted by permission from *TIMSS Mathematics Items, Released Set for Population 1 (Primary)* prepared by the IEA TIMSS International Study Center at Boston College (1994) and copyrighted by IEA, Amsterdam, the Netherlands.

Figure 9–13 is reprinted with permission from "Measurements" by Lindquist and Kouba in *Results from the Fourth Mathematics Assessment of the National Assessment of Education Progress* edited by N.M. Lindquist. Copyright © 1989 by the National Council of Teachers of Mathematics.

The standards in Figure 4–1 are reprinted by permission from *Grade Two Curriculum Expectations*, Glenview School District 34, Glenview, IL; *Illinois Learning Standards*, 1997, Illinois State Board of Education, Springfield, IL; and *National Science Education Standards*, 1996, National Academy of Sciences, Washington, DC. Examples of how local objectives can be written to align with the Illinois Learning Standards are those of the author and the Illinois State Board of Education. Permission to reprint the excerpts does not constitute endorsement of those examples of alignment.

"Math Inquiry of Arrays" in Chapter Four was developed and used by permission of Diane Czerwinski.

Figure 10–4 is reprinted by permission from *Best Practice in School Science Education, 2001* by the Science Review Committee, Glenview School District 34, Glenview, IL.

Library of Congress Cataloging-in-Publication Data
Heuser, Daniel
    Reworking the workshop : math and science reform in the primary grades / Daniel Heuser.
        p.   cm.
    Includes bibliographical references.
    ISBN 0-325-00433-1
    1. Mathematics—Study and teaching (Primary)—United States.
2. Science—Study and teaching (Primary)—United States.   I. Title.
QA135.6 .H48 2002
372.7—dc21   2002004345

Editor: Victoria Merecki
Production coordinator: Elizabeth Valway
Production service: Lisa S. Garboski, bookworks
Cover design: Joni Doherty
Typesetter: Publishers' Design and Production Services, Inc.
Manufacturing: Steve Bernier

Printed in the United States of America on acid-free paper
06   05   04   03   02   RRD   1   2   3   4   5

To Jane, Katie, and Benji

Contents

. . . .

Acknowledgments

. . . .

Many people helped make this book possible. First and foremost, my family. My wife Jane put up with a lot of extra baby-sitting and conversations over the dinner table about "the pedogological implications of research." Her constant love and support were the most valuable contribution to this book. My daughter Katie and son Benji provided their usual endless amusement as well as the motivation to make their educational future brighter. I also want to thank my parents for their interest in the project and for their perspectives on education.

The greatest influence on my work has been that of the late Gerald Foster of DePaul University. Like all great teachers, Jerry called on me to question my most basic beliefs. Such great knowledge rarely comes with such a warm, helpful, and unpretentious personality, and Jerry will be missed. Also at DePaul, thanks go to Roxanne Owens and the interns at the DePaul Glenview Clinical Model Program for their help in collecting the Math Workshop Project data. Terri Pigott, a friend at Loyola University, helped analyze this data.

Glenview Public School District 34 is the ideal environment in which to write about great education. The children, parents, teachers, and administrators of Glenview have supported this effort from the beginning. Several individuals stand out. Allison Straussman, my teaching partner, provided insight, feedback, and a laboratory to refine my ideas. The staff and principals at both Lyon and Westbrook schools gave tremendous assistance. I would especially like to recognize Charlotte Miciek, Pat Neudecker, Frances Brown, Debbie Shefren, Dave Work, Debbie Gurvitz, Diane Bilcer, and Becky Dean. Special thanks go to Jan Kirch and Jean Conyers for not shying away from what is right for children.

At Heinemann, Victoria Merecki was a model editor. Her professionalism and knowledge are much appreciated.

Finally, I want to thank the scholars: Constance Kamii (University of Alabama), Darrell G. Phillips (University of Iowa, retired), Dale R. Phillips (Grant Wood Area Education Agency, retired), and Thomas P. Carpenter (University of Wisconsin). Their writings challenged and inspired me; these individuals truly answered the call for wisdom and understanding.

Introduction

· ·

John Glenn's return to Earth on a sparkling clear morning in 1962 was a watershed event in how Americans viewed mathematics and science. Glenn had just completed the first American orbit around Earth, and as his capsule plunged into the Atlantic, the power of math and science was rarely so visible, and their promise so bright. Forty years later, however, the now former astronaut and senator Glenn delivered a far different message: the mathematical and scientific needs of our society were evolving so quickly that our educational system was becoming obsolete. Glenn's committee released a report—aptly titled "Before It's Too Late"—that chronicled the desperate need for math and science educational reform.

It is teachers like you and me who are the pioneers in this reform. Policy makers and researchers also play a part, but it is teachers who bring ideas to life as we go about our day-to-day calling: teaching children.

This book is about teaching math and science better. It addresses three big ideas:

Some ways of teaching are better than others. Research and experience show that certain teaching techniques are clearly superior in helping children construct math and science understanding.

Good math and science teaching looks a lot like good reading and writing teaching. Reform teaching techniques are very different than those traditionally used in math and science instruction. Instead, they mirror what happens in successful literacy classrooms, especially those using reading or writing workshops. Connecting math and science workshops to literacy workshops can help make changes in teaching easier.

Math and science workshops combine reform teaching techniques with a familiar, successful lesson format. Reworking the workshop means adapting the workshop elements—curricula, minilessons, activities, reflection, assessment—to bring sustainable and significant changes into your math and science teaching.

This book is a part of a larger movement to reform math and science education. The goal is that children will no longer see these subjects as collections of things to memorize. Instead, math and science will be the natural, understandable, and enjoyable means to make sense of the world around them.

About This Book

· ·

When you were a student, how did your teachers get you to understand and love math and science? Or did they? In Chapter 1, I explain my experiences learning (and not learning) math and science. If you change a few details, you may very well find that this chapter is about you. What can be learned from our lives both in and out of school that will help us be better teachers?

With the personal connections gained in the first chapter, Chapter 2 describes the research base of effective math and science teaching. It is important to know that the elements found in great learning experiences outside of school are supported by a broad, extensive body of educational research. This chapter will help you develop your own strong understanding of effective teaching—and be able to use it to inform fellow teachers, administrators, and parents.

Chapter 3 illustrates the many similarities between good math and science instruction and good literacy instruction. Seeing the philosophical and practical connection with something that you may already use successfully (the reading and writing workshop) will ease the transition toward teaching math and science through the workshop.

The next seven chapters show specifically how to plan and teach workshops. Three sequences of workshops are detailed throughout these chapters. Watching these workshop sequences play out, from the beginning planning to final assessments, will help you use workshops in your classroom. Specific classroom examples, hints for troubleshooting, and results of action research help illustrate each topic.

Chapter 4 shows how to plan your math and science program. Given the research and the national standards, what is appropriate curriculum? What activities best help children learn? How do workshops fit into the overall math and science program?

Chapter 5 provides background information and details on how to teach three specific workshop sequences: addition facts, life cycles, and the range of activities that make the student-directed workshop.

Chapter 6 explains how to conduct a workshop minilesson. Topics include motivating children to learn and setting expectations. Several sample minilessons are described in detail, including some from the three sample workshops sequences.

Chapter 7 illustrates how to conduct a successful activity period. What is the teacher's role as students work? How does hands-on activity affect children's mental development? Activity periods from the sample sequences and other workshops are described.

Chapter 8 shows how to foster thoughtful reflection. The use of pair shares, discussion, and learning logs will be detailed, along with how to nurture a classroom climate that encourages reflection. Reflection in workshops, including the three sample workshops sequences, is discussed.

Chapter 9 explains the use of various assessments. This chapter shows how to use assessments such as interviews, embedded assessments, artifact assessments, and questions adapted from standardized tests. Additionally, this chapter shows the crucial role that assessment plays in planning future workshops. Multiple ways of assessing student understanding are illustrated from the sample workshop sequences.

Chapter 10 addresses the issues that teachers are likely to face as they reform their math and science teaching. Proactive strategies for informing colleagues and the wider community about workshops are described, including newsletters, parent nights, and communication with your principal.

Math, Science, and Me

This chapter is about three of my favorite subjects: math, science, and me. You may also find that it is about you. Many other teachers have related experiences and feelings similar to the ones I am about to describe. You, too, may find yourself thinking about comparable events from your math and science life. Like some of these other teachers, you may not have especially liked these subjects when you were a child, and they may not be your favorite parts of the school day now. Or perhaps you are more like those who are generally positive about math and science. Because you are reading this book, however, you probably have one thought in common with all of these teachers: is there a better way to teach math and science?

MATH AND SCIENCE WORKSHOPS

Math and science workshops are a better way. Workshop teaching looks considerably different than traditional math and science education. Results are also different: children develop a deep understanding of these subjects. Math and science are no longer viewed as just facts and procedures to memorize. Instead, they become rational, useful ways to solve real problems and satisfy personal curiosities.

Teachers also find advantages to using workshops in math and science. Workshops may already be a successful part of how you teach reading and writing. The transition to constructivist, research-based teaching can be difficult, and this familiarity with the workshop may ease the move away from more traditional instruction.

REFLECTING ON MY MATH AND SCIENCE EXPERIENCES

For now, however, this chapter is about math, science, and me. As a first-year teacher of 25 first graders, I had one simple goal. It seemed pretty reasonable at the time: I wanted to teach math and science lessons that didn't make everyone involved feel horrible. I was tired of fighting to keep the kids' attention and feeding them answers that they really didn't understand. My students were tired of hearing me talk and memorizing things that were of little interest to them. No matter what I tried that year, math and science instruction were frustrating puzzles.

As I struggled to improve my math and science teaching, there was one part of the day that always went well. During writing workshop, children were focused, worked hard, and enjoyed the time. They actually complained on those days that we didn't have it. More importantly, they grew as writers far out of proportion to the amount of traditional teaching I did. As I marveled at their progress, I was faced with another puzzle—what makes the writing workshop so good?

True to the title of this chapter, I discovered the solution to all these puzzles as I thought about . . . myself! I examined more than my time as a teacher. I reflected on what I did outside of school, too, as well as searching far into the past to when I was a child. Were there some common elements to be found in those situations in which I learned a lot of math and science? What seems to spark great learning in these disciplines? Eventually, four common elements emerged: hands-on activity, choice, reflection, and problem solving and inquiry. To explain, let me take you back a few years.

A Student's Perspective

My first memories of math and science are from elementary school. During second grade, my friend and I found a perfect laboratory for hands-on science problem solving and inquiry: a small drainage ditch at the edge of our schoolyard. In the shallow water we discovered snails and a crayfish. Over the next several weeks we spent our recesses at that ditch, examining the animals or redesigning their home by moving around rocks and sticks. One day, to our horror, we found a colorful streak of oil floating on top of the water. We immediately went to work, fishing out the endangered animals and putting them into an old bucket. During the frantic rescue operation our newly found environmental activism emerged as disgust: How could anyone do this? Doesn't anyone care about the Earth?

We hauled the bucket to our teacher and tried to convince her that it contained potentially great classroom pets. Sensing our earnest environmental awareness (or maybe smelling the bucket's contents beginning to ripen under the early summer sun), she suggested that instead we try to clean up the water. Grabbing handfuls of paper towels, we ran back and began to blot up the oil from the surface. We returned the animals to their newly cleansed home, and, for a while at least, our fragile paradise was again safe.

Why did I remember this incident? All four learning elements were in place to make this experience so lasting. I was involved with *hands-on activity*, experiencing nature directly, not through reading a book or listening to someone else. Being there was my *choice*; many of the other kids chose the kickball field. The time spent discussing and debating with my friend was the *reflection* that helped me internalize the whole episode. I was also involved in *problem-solving* and *inquiry*, because we were concerned and curious about the oil spill and because my friend and I chose our own solution method to clean the water.[1]

Seven years later I was sitting in a high school biology course, studying the same crayfish and snails that had intrigued me as an 8-year-old. The focus of the class was animal classification. My teacher was an excellent speaker, very dynamic and clearly passionate about living things. Specimens in formaldehyde filled the shelves, and a fish tank and plants shared the windowsills. Every class period he described the characteristics of another group, and my afternoons were filled with talk of pseudopodia, phyla, invertebrates, and Monera. As that long winter wore on, however, it became apparent that we wouldn't actually get to *see* snails or crayfish or any of these creatures whose family trees we were busy scribbling into notebooks. There were no labs or field trips, and the specimens on the shelves never left the shelves.

Instruction in that class was by lecture, video, and textbooks only; there was no hands-on interaction with any of the animals. Seeing a picture of a sea anemone or a pig didn't nearly have the same effect as feeling the anemone's sharp spines or observing that, yes, pigs really do have hair! Nor did I know *why* I was studying animal classification. Because it wasn't part of an inquiry to satisfy one of my own questions, it was difficult to maintain interest.

Another problem was that this teacher didn't encourage reflection. In a sense it was unfortunate that he was such a dynamic speaker; talk flowed one way—from him to us—and there was little opportunity to question him or discuss among ourselves. This arrangement invited memorization rather than knowledge building. I learned specific facts: for example, crayfish are crustaceans and crustaceans have exoskeletons. But I didn't come away with any kind of lasting understanding of the animal kingdom, and I certainly didn't feel any of the passion that I had found in that school drainage ditch. Rather, I became convinced that science was all about memorizing and then repeating it on a test.

I held a similar view of high school math. Although I had always done well in grade school math, I quickly learned in high school that geometry, algebra, and trigonometry were not meant to be understood.

---

1. Problem solving and inquiry are closely related process; their similarities and differences are described in Chapter 4.

As with science, teaching generally didn't include hands-on activities. Choice and reflection were not on the instructional menu, and there was little opportunity for me to solve problems in ways that made sense to me or to inquire into my own interests. Instead, we did endless strings of exercises, using the method that the teacher modeled. I memorized how to solve a quadratic equation, to solve for x and y, and to calculate the area of an oval. And I repeated all of it on a test. But this knowledge was so bound to the textbook and the classroom that I forgot what I had learned as soon as I stepped out of the school context.

As I moved into university-level math and science, the limited hands-on instruction, choice, reflection, and problem solving and inquiry found in high school virtually disappeared. I took some fairly advanced courses as a premedical student. Yet I managed to come away with almost no understanding of the purpose or spirit of any of these subjects. As in high school, I could complete most of the exercises. But without these four elements, I was unable to tie the abstract ideas into reality. I could memorize and repeat, but I really didn't "get it."

Even so, I got As and Bs in all of these courses. How did I do so well without understanding? My roommate, John, and I used a time-honored approach. Success always came down to the night before the test. Fueled by a fatalistic bravado and multiple cups of coffee, we stayed up all night going over our notes. We memorized facts in sharp detail. We would quiz each other: "What year did Darwin set sail on the *Beagle?*" John would ask. "1831!" I would snap back. "And that's on the bottom of page 239!" In the morning we would walk quickly to class, careful not to talk to anyone lest we lose some important knowledge. After the test we would take a well-deserved nap, and, as if by magic, we would wake to find that we had forgotten everything we had "learned."

The teachers didn't help much either. Reflection, especially, was a vital element missing from these courses. Discussion of what calculus was actually used for—building bridges? predicting the weather? planning finances?—may have helped tie the abstract symbols to their real-life applications. Reflecting on how the types of problems were related would have helped us connect to what we already knew. Instead, my calculus instructor spent the entire class with his back to us, frantically writing problem after problem on the blackboard.

After 2 years of premed I had taken six advanced math and science courses. Very little from these courses made any sense to me. I had few hands-on experiences. Choice of what to study or how to solve the problems was not allowed. Reflection was not encouraged. Meaningful problem solving or inquiry rarely took place. Perhaps it is no surprise that I got out of my formal science major at the end of that year. Like many students, over my school career I had developed two parallel belief systems about math and science. *School math and science* was to be memorized until after the test. *Real-life math and science,* on the other hand, was useful outside of school for building things, balancing my checkbook, and making sense of baseball statistics.

Real-Life Math and Science

And arguing with my sister. The earliest real-life application of math or science I can remember took place on the way home from Sunday School in the back seat of the family car. My sister announced that 10 tens were 100. She was in second grade, 2 years older than I was, and often seemed to make similarly outrageous claims to pass the time on the ride home. I found the idea that ten tens equaled 100 extremely laughable. One hundred was a BIG number. It was the ceiling on my counting ability. I had been astonished when I found out earlier that my dad made more than $100 a year. One hundred was so big, I thought, you must be able to fit at least 20 or 30 tens into it.

She tried to prove her point by counting by tens. "10, 20, . . ., 100. See, 10 tens!"

"That's not how you count!" I informed her. "You start with one, like this: "1, 2, 3, ..."

Saddened by my ignorance, she started adding tens: "10 + 10 = 20, 20 + 10 = 30, ..." Not being able to add, this argument did not sway me. Even when my parents agreed with my sister from the front seat, I would not believe it.

"Here," I said, grabbing my Sunday School pamphlet and a pencil. "I'll show you." In the margin I started tallying lines: / / / / / / / / / /. When I drew 100 lines I began to circle groups of 10 lines. It was not until I circled the eighth group that I started to realize that my sister could be right. When I was done, I stared in amazement at my paper—100 lines in 10 groups of 10. Ten tens in 100!

In this episode, Are there really only 10 tens in 100? was my inquiry question. Using my chosen method, I proved to myself what my sister and parents couldn't prove with their more advanced arguments. Their evidence couldn't convince me because it was beyond my understanding. It was only when I chose a technique that made sense to me that I could construct this new knowledge of 100.

The most powerful adult example of using math and science outside of school was the time my wife and I built a backyard pond. This project is a clear example of the amazing effectiveness of hands-on, choice-driven, and reflective problem solving and inquiry. Planning and building the pond required considerable math and science. For weeks we tangled with mapping sun patterns, digging 60° sides, maintaining proper pH, and identifying native plants and animals.

Over the years, watching the pond change and mature has led to a deep understanding of this tiny part of the natural world. I know the concept of life cycles in vivid detail by observing generations of frogs from egg to adult and once, armed with a flashlight, by seeing larva wriggle to the pond surface and transform into a mosquito. Cause and effect shows its many outlets in the pond: how duckweed covers the surface as the temperature rises, how fish die off as the pond becomes overpopulated, and how the ecosystem slows as the autumn shadows fall across the yard.

My understanding of the pond increases as I make further inquiries into what interests me. For example, in order to have a self-sustaining frog population, we experimented with several species to determine which would breed. One summer we were finally rewarded with mating choruses. For several weeks a lonely *buum, buum, buum* floated through the night air into my daughter's bedroom window. Soon I found clumps of eggs. Although they did develop into tiny tadpoles, almost all of them were deformed in some way, and none survived. Whether or not this was related to the worldwide problem in amphibian populations is unclear. This is one ongoing inquiry from which I can expand my knowledge of the world of science.

Learning real-world math and science is considerably different than learning in school situation. In most cases I've learned by doing: counting, measuring, experimenting, and observing. Tangling with real-world math and science is often by my own choice, and I'm free to solve problems any way I can. Outside of schools these subjects often happen in a social context, where articulating and debating ideas are natural ways to reflect. Additionally, I use math and science in real life to inquire about curiosities and to solve problems that are important to me. I have no doubt that my real-world math and science experiences have led to greater understanding of these subjects than have my years in school science and math. And much of that stems from the natural presence of hands-on activities, choice, reflection, and problem solving and inquiry found in math and science outside of the traditional classroom.

Lessons of a Preworkshop Teacher

A third source of reflection for me was my first few years as a teacher, before I began using workshops. During that time, I taught math and science in a way that resembled more of how I was taught in school than how I learned in real life. These experiences further convinced me of the effectiveness of hands-on learning, choice, reflection, and problem solving and inquiry.

To teach multidigit subtraction before the workshop, I usually presented several different problem-solving methods. These included counting down, subtracting the tens first and then the ones, and the standard adult algorithm. After I solved a few sample problems I would send the children to practice what they learned on their worksheets. I dreaded the next fifteen minutes. Invariably 10 hands would shoot up to request help—that is, if the students didn't surround me and start tapping my shoulder. The meekest children of the group would just sit and stare at the paper. The more aggressive would "borrow" answers from the brightest kids. And then there were those hovering around me announcing, "I'm done! What do I do now?"

Even when a child was genuinely motivated to solve a problem and I was free to help, the inadequacies of my memorize-and-repeat teaching were clear. For example, one girl was trying to subtract 25 from 50. "You can take 2 away from 5, and that gets you 3. But how can you take 5 from 0? Is the answer just 3?"

Hoping to put some meaning to the numbers, I ventured, "Look at it this way. What if you had 50¢ and you took away one quarter? How much would you have then?"

"That easy: 25¢," she quickly replied. "But how can I solve *this* problem?"

It was clear from her speculation that the answer may be three that this child was not thinking of the true sense of this problem. It was also clear that she was focusing so much on applying the rules that I had just explained that she could not see the connection to money. This may not have been the case had she been given the choice to solve the problem in a way that made sense to her. Had she been encouraged to choose, she might have counted out manipulatives or drawn pictures to represent the two numbers or maybe used coins. But she couldn't because I didn't think to put choice on the menu in my classroom. And with the manipulatives safely stored away in my closet until another hands-on activity was scheduled, choice was just not convenient.

My preworkshop science instruction was also lacking. My team and I taught an inquiry unit on water. Unfortunately for the children, glaciers and oceans are a long way from Chicago, and rather than get their hands wet conducting hands-on investigations, they had to read books to find the answers to their questions.

One of the students, Jake, taught me another weakness of hands-off learning. Jake's question was, "What makes ocean water salty?" His teacher asked him for a hypothesis, and Jake replied that there must be someone somewhere shoveling salt into the ocean. Like most children, Jake probably at one time had spooned salt into a glass of water and tasted it. So if a spoonful makes a glass salty, according to Jake's first-grade logic, it must take a shovel to make the ocean salty.

The teacher acknowledged his idea and sent him off to find the answer in a book. Jake was a very good reader and soon found another explanation in a reference book. There are rocks in the ocean, the book noted, that contain salt. The water continually dissolves this salt, which makes the water salty. Jake reported this to his teacher.

The next day, the teacher asked each person to prepare a poster showing his or her question about the ocean and the answer to that question. Jake's poster vividly illustrates the resiliency of children's naive ideas about the world. On the top he wrote in big letters "What makes ocean water salty?" Underneath was a picture of a man, feverishly shoveling mounds of salt into the ocean water (Figure 1–1).

Jake's "finding the answer" in a book did little to change his original perception of the natural world. I have had many similar experiences with children who are forced to rely on books, videos, or even my own words to form new understandings. More often than not, these secondary experiences fail to alter ideas that children have previously formed through hands-on experiences.

My math and science teaching before I started to use the workshop was a mixed bag. Sometimes my students learned a lot and had very positive feeling toward those subjects. More often they saw them as lessons in memorization that didn't really make sense to them. In almost all the cases, however, the presence or absence of hands-on instruction, choice, reflection, and problem solving and inquiry usually made the difference between understanding and memorizing. How can these four elements be combined into a practical teaching format that promotes students' achievement in math and science? One answer is the workshop.

And Now: Teaching with the Workshop

Teaching math and science through the workshop has been immeasurably more successful than my preworkshop lessons. Two examples of workshops—one math and one science—show the results of systematically incorporating hands-on instruction, choice, reflection, and problem solving and inquiry into each lesson. With these four elements, workshops resemble my best math and science moments rather than my worst.

Math Workshop   The area of a shape always seemed an easy concept to teach. Prior to using math workshops, I taught area traditionally, through rote learning. My students' workbooks had many two-dimensional shapes on which graph paper was superimposed. I told them to count the squares, and there you would have the area. Simple! I noticed, however, that if no little squares were provided, children had no idea how to find the area. They had memorized a simple procedure and were unable to extend it outside of very specific situations.

FIGURE 1–1   A student's resilient belief about what makes the ocean salty.

Teaching area through the workshop is considerably different. To begin the workshop minilesson, I hold up two differently shaped papers of similar size, shape A and shape B (Figure 1–2). I ask a question—Which one do you think has the most space inside?—and the process of problem solving begins. Students excitedly share with each other which one they think is bigger. After taking a few opinions, I ask how they could find out for sure. Again students turn to each other and share their ideas.

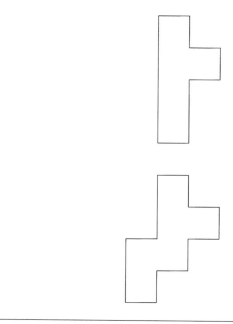

FIGURE 1–2   What shape has the most space inside?

These first and second graders have never been formally taught area, but each has some informal experience comparing shapes, and I want to draw on that prior knowledge to solve a new problem.

Now that each child seems interested in the problem, the rest of the minilesson explains what they will be doing during the activity period. For this activity period, the children are in pairs. I often have children in small groups or pairs because the discussion that often ensues facilitates reflection. Each pair of children is given its own sets of shapes A and B, and the sets all slightly different. I make it clear that each person is to solve the problem in whatever way he or she chooses. I also show them a variety of tools they can use, including base 10 blocks, colored square tiles, and dominoes, although they don't have to use any of them and can use another tool if they want. Then, they begin the activity period.

The room is a flurry of hands-on activity. Each pair is actively involved in solving the problem. It surprises me that many of the children do not use the tools that I provided. Some students are filling the shapes with these objects, but others have folded one of the shapes to try and match the second shape. Still others have cut one shape in parts and laid them atop the other shape. This is another example of the great benefits of allowing children choice: by not requiring them to use what makes sense to *me*, they can invent ways to solve problems that seem logical to *them*. Each of their methods could be a way to compare area.

As the children work, I confer with some of the students. As in writing workshops, conferences are simply a time in which I can see students' work, listen to their explanations, and challenge their thinking. Each conference is individualized, based upon students' abilities and what they are doing. As usual, in this workshop I am finding myself doing far more questioning of students than answering students' questions. The questions I'm asking today—How do you know shape A is bigger? How did you figure that out? What's another way you can solve this?—are designed to get children to verbalize their thoughts and defend and clarify their problem-solving strategies. This process of using language to define thought is key to helping children reflect on their thinking. I also use questioning to point out inconsistencies or faulty reasoning in the way some children are solving the problem.

After the activity period is time for reflection. Children have already reflected through the conversations in the activity period. However, the reflection period helps ensure the systematic and deep reflection needed to turn hands-on experiences into knowledge.

To begin the reflection, I ask students to share their problem-solving strategies with another person other than their partner. These "pair shares" are a useful way to help even the quietest children think about and communicate their ideas. I then ask for volunteers to share their strategies with the whole class. As Harriet and Tom show how they cut their shape A into squares and then laid the squares on top of shape B, Elaine challenges them. "Now you know which one *was* bigger, but you ruined Shape A. You should trace shape A onto another piece of paper, then cut *that* one into squares. That way you won't ruin it." Paul suggests, "Why don't you just use tiles? Then you don't have to do any cutting or tracing." This conversation continues as children argue for or against each method. In the end, we are beginning to sort out which methods are best, and the children are aware that there is more than one way to solve this problem.

The reflection period finishes with students writing in their learning logs. The children respond to two questions. The first is, How did you find which shape had the larger area? Since the children have had several opportunities to rehearse explaining their procedure, I am curious how completely they can communicate it in writing. The second prompt is to encourage children to apply their understandings of area to a new situation: How could you find the area of the floor of your room at home using things that are in your room? Because floors can't be cut or folded and students probably don't have base ten blocks or color tiles at home, this question doesn't allow for rote application of memorized procedures. Instead, it offers an opportunity for students to extend their understanding and gives me an assessment to chart their progress.

Although I have broadly planned out the next several math workshops on area, what I learned during this workshop helps me refine my plan. I know that each new experience needs to be thoughtfully connected to what they have already done. Using their log responses as well as the mental notes I took during the activity period, I have a better idea of what tomorrow's workshop should look like in order to dispel misconceptions and deepen their understanding of area.

Science Workshop   Before I taught science through workshops, my teaching of life cycles mirrored my very worst science experiences. Living things don't like to be touched (tadpoles and butterflies), grow painfully slow (plants), and unexpectedly die (all of them). With hands-on activities so difficult, my students spent a lot of time reading, watching videos, and listening to me talk.

Teaching with workshops changed this situation, because knowing I had to use hands-on activity, choice, reflection, and problem solving and inquiry in nearly every lesson forced me to reconsider my objectives. Rather than focusing on a stilted objective–students shall know the life cycles of the bean and frog–I aimed on helping students develop an understanding of how living things change over time. This broader objective allowed me to incorporate parts of other second-grade units—conservation, environment, water, and soil—to make the unit a more coherent whole.

The first workshops in the unit focused on the question, What is in a pond? We spent a whole morning at a local nature center taking water samples, observing and sketching the numerous plants and animals in the water. We walked around the edges and discussed what we saw there. The center also had a nature museum and aquarium, where we observed pond animals more closely. Activities on the days following the trip were meant to capitalize on their excitement and draw on previous knowledge that children had of wetlands. Through discussion, drawing, and writing, children were encouraged to link past experiences with what they learned on the field trip.

This trip to the pond was an important stimulus for our next workshop: an inquiry into how tadpoles develop. I have found from experience that children are more curious and ask better questions after experiencing phenomena firsthand. Without opening hands-on activities, children often ask questions that are vague and untestable, such as What is photosynthesis? or Why is it called photosynthesis?

To capitalize on students' interest I suggest that we try to make our own pond in the classroom, modeled after some of the exhibits we saw in the nature center aquarium. We use an aquarium sitting empty since the beginning of the year. We debate what animals we could include in our minihabitat. Among the animals we saw at the nature center were frogs, and the children decide that they would like a frog in our miniature pond.

These creatures became the focus of the next workshop in this unit. The frog eggs arrive in the mail, and over the next couple days we watch them hatch. Students record their observations in a series of observation journal drawings. Inquiry is a teaching method that relies on students' questions, so I keep my ears open while they watch and draw the tadpoles. Questions that focus on the tadpoles' environment, such as Why did that one die? How long until they are frogs? What do they eat? dominate the children's curiosity. I summarize their thoughts by posing the umbrella question, What environment is best for tadpoles to develop into frogs?

Students volunteer their ideas to what kind of conditions would be best for the tadpoles as I record them on the board. Many suggestions show that children are relating tadpole health to human health. For example, some suggest feeding them cheese or giving them a lot of exercise. Other children bring in their experiences at the nature center. One girl, for example, feels that the tadpoles should be fed duckweed because the nature center pond was covered with duckweed.

When the list is complete, I cross out those conditions that would probably harm the animals or are impractical. Gone are feeding the tadpoles hamburger or putting them in my bathtub at home. From the remaining suggestions, each child is required to pick one that they think would be the best for helping the tadpoles to develop. In their learning logs they record their prediction, why they think that condition will help the tadpoles, and how they can test their idea.

That evening I put children who had similar but conflicting predictions into groups of three or four. For example, two children who thought that muddy water would be best were grouped with two who felt clean water was better. This avoids a common phenomenon in science, that of children ignoring evidence that contradicts their initial beliefs. Pointing out to children that their beliefs are inaccurate is a powerful tool to help children change their concepts. I know one of the pairs of children will gladly point out the other's mistakes as they reflect on their results.

The next day each group meets and shares their predictions as well as their plans about how to test them. Children are expected to decide how their group should conduct the experiment. For example, one child in the muddy/clean-water group wants to put dirt on one side of a container only and see to what side the tadpoles swim. Another child in the group points out that it will not matter if one side does not have dirt, because the water will all get muddy anyway. By presenting and debating the plans, the children finally negotiate a hybrid plan with which each child is comfortable. They then draw up their ideas more formally, including a list of materials and a detailed drawing of the materials in place. Each of the plans is then presented to the whole class.

The next day we put our plans into action. Students prepare and place their tadpole tanks. Containers are large or small, water is muddy or clean, and the containers are placed in the sun or shade, all based on the investigation the students choose to pursue. Similarly, some tadpoles are fed lettuce or pellets, and others receive fruit or duckweed.

Over the next few weeks, students stop by to check on the containers. Interest is high as they hover over "their" tadpoles. They put a tally on the container label whenever a tadpole dies. These events often are the topic of peer talk without my prompting.

Once a week the habitats are observed as part of the ongoing reflection in this workshop. The groups sit with their containers in front of them and share what they see. This usually leads to debate. In the muddy/clean-water group, for example, one boy explains that muddy water must be better because those tadpoles look bigger. Another child disagrees, arguing that one tadpole had died in the muddy water but none in the clean. Overhearing their discussion, I encourage them to modify what "best for tadpoles" really meant—should it be growth or mortality rate? These reflection periods conclude with each group presenting their findings to the whole group and every child recording what he or she learned in a learning log. Through this extensive reflection, students articulate and refine explanations in order to incorporate the hands-on work into their current understanding.

Before teaching with workshops, I would often conclude lessons by telling the children what they had learned. For example, I would announce that their inquiries showed that animals develop through their life cycle best when they are in conditions similar to their natural habitat. This now seems foolish, for two reasons. First, if telling children the answer gave them understanding, I wouldn't have to bother with the messy hands-on activities. I could just lecture them on the given concept, and they would know it. I know from experience, however, that most children don't learn very well from just listening to the teacher. Second, whatever I understand to be the right conclusion is synthesized through my adult intellect from a lifetime of experiences. Children will not necessarily understand or agree with it. By hearing my adult conclusion, children will just memorize what I say at the expense of their hard-earned knowledge gained from inquiry. This will discourage them from trusting their own senses and reinforce the idea that science has to be memorized because it doesn't really make sense.

Workshops are structured to allow children the freedom to learn what they are ready to learn. Because What environment is best for tadpoles to develop into frogs? is a fairly broad question, there is considerable room for children to draw different conclusions. They understand these conclusions because they are based on their experiences and intellect. I also know that most of these conclusions are just as "right" as mine are, and act as a temporary understanding to be revised through later science learning.

With that in mind, I conclude this 3-week-long inquiry by asking children to reflect on what they learned from the tadpole experiments. The children's learning log responses demonstrate an astounding depth and variety of knowledge. My special education student is able to accurately report on his group's experiment: the pellets are good food. Other students incorporate the results of several experiments. For example, one girl concludes that a larger home of muddy water and pellets as food would be the best environment. That some children acknowledge their beliefs have changed based on their inquiry is reflected in the following quotation: "I changed my mind. I thought that duckweed would be better, but now I think that pellets are better. Those tadpoles are bigger. I wonder if duckweed *and* pellets would be the best?"

Other students synthesize the results of all the experiments with their observations of the nature center pond and their previous knowledge about animals: Tadpoles grow quicker in shady water, because in the pond the water was totally covered with duckweed. The pellets have dead fish and plants in them, which would be the tadpoles' food in a pond. The best way to grow a frog would be to make the tadpole home the same as a pond, because that's where they normally live. That's probably why they set up zoo cages with rocks and trees to look like where the animal actually lives. Reflections such as these are very similar to the scientifically accepted adult conclusion. I find this kind of remarkable learning far more often since my move away from traditional instruction to science workshops.

CONCLUSION

The systematic use of hands-on instruction, choice, reflection, and problem solving and inquiry in workshops results in some powerful learning. Children clearly demonstrate a deep and flexible understanding of these important math and science concepts. Students are motivated to learn, and they enjoy workshop time. From my perspective, workshops are that better way to teach math and science.

Writing this chapter was an excellent exercise in self-awareness. In almost every situation I considered, I reached the same conclusions about how I learn best. Additionally, I have noticed that my students not only are ready to learn in a hands-on, choice-driven, reflective, problem-solving and inquiry model of teaching, but thrive under these conditions. The fact that so many teachers have voiced their agreement has also been gratifying—I know that I'm not the only one who sees it this way.

But this book isn't about adults sharing feelings; it's about teaching kids. And although it would be tempting to base our teaching practices on our common experiences—we benefit from these elements, so let's teach children in the same way—doing so is ultimately narrow-minded. Everyone is not like me or like you. To truly benefit children requires moving beyond fuzzy childhood memories and adult revelations to ask the broader question, Will a model of teaching that systematically uses hands-on instruction, choice, reflection, and problem solving and inquiry help children gain the understanding needed to thrive in this new century? To answer that, we have to momentarily turn away from ourselves and look at the research.

2

The Research Base

I magine sitting in your doctor's office as she writes out a prescription. "You know," she says, "I've been prescribing this same medicine for 25 years!" "What?" you reply. "You mean after 25 years of medical research, hundreds of scientifically controlled clinical studies, millions of research dollars spent, and some of the brightest minds dedicating their lives toward bettering treatment, there's been absolutely *no improvement?*"

In countless classrooms across the country, this same scenario is the norm—teachers are teaching math and science in essentially the same way that it was taught to them when they were students. The task of taking on teaching methods that are unfamiliar is notoriously difficult, but not making these changes has serious consequences. American students, as they are presently taught, lack an understanding of the essential concepts, don't like or trust math and science, and stop taking these courses as soon as they can (Schmidt, McKnight, and Raizen 1997; U.S. Department of Education 2000; National Research Council 1989).

Could the four elements that I believed to promote understanding— hands-on instruction, choice, reflection, and problem solving and inquiry—help address these issues? As I began to look at the research, I wondered if I could find enough studies that examined these elements. As it turns out, the more I searched, the harder it was to find research that *didn't* address them. Each element has been the subject of considerable study. This is important for two reasons. First, it may illustrate that educational research can be seen to be a comprehensible whole to guide teaching, rather than a dizzying number of unrelated studies. Additionally, because the findings on math education research overlap greatly with research on science education, understanding of the four learning elements can be used to modify teaching practices in both subjects simultaneously.

As I finished my review, I was startled at the consistency between my experiences and the research. The evidence supporting these four elements is considerable and forceful. This chapter alone relates the results of dozens of studies conducted over a quarter of a century. Hands-on instruction, choice, reflection, and problem solving and inquiry all appear to greatly increase young children's achievement in math and science.

WHY RESEARCH?

Even though many people already use research in other areas of their lives, such as when deciding what kind of car to buy or medicine to take, educational research can be difficult to use. Between the ungainly style of written research, the daunting scope and number of studies, and the belief that formal research is not always applicable to the real world of teaching, research is not universally read or discussed by teachers.

Why should teachers care about research? Several processes influence how we teach. Many teachers use their first-hand experiences—what seems to work or not to work—to guide their teaching practices and philosophy. At other times, feedback from outside sources help validate ideas. It helps when trusted colleagues, a principal, or an author share a teacher's vision.

Some teachers have gone one step further, melding their experiences and the beliefs of those they trust with the results of research. Research helps to minimize the bias of personal observations and the potential inaccuracies of feedback from others. Research is a systematic, scientifically controlled way of observing, collecting, organizing, and examining genuine phenomena to show what actually does and doesn't help children learn. This makes research one of the most powerful tools teachers have to guide their professional growth.

Research can also be used to educate and influence parents, administrators, and other members of the community. Because research-based teaching often looks different than traditional instruction, it is essential for teachers to communicate the rationale behind the differences. Using the results of research can both guide and lend validity to teaching practices.

HOW CHILDREN LEARN

To know how to teach better, you have to know how children learn. The great majority of educational researchers and theorists believe that one theory—constructivism—is the most accurate model of children's learning (Battista 1999). Constructivism states that knowledge cannot be given to children, but rather that the child has to create his own knowledge based on experiences. Physical experiences, such as working with clay or observing insects, and social experiences, such as class discussions and reading books, offer opportunities for building knowledge. To transform these experiences into knowledge, children need to abstract and reflect upon their experiences. This preliminary knowledge is likely to change as more experiences, mental development, and reflection occur. In this way knowledge is continually evolving.

To Memorize or Understand?

$2 + 2 = 4$. Knowing this fact is almost a coming-of-age event in the kindergarten classroom. A beaming "I know how to plus!" or a nonchalant "Everyone knows that!" often follow when young children repeat this equation. Of course, knowing that $2 + 2 = 4$ does not equal a deep and flexible understanding of addition. Rather, it could be an example of a child reciting a memorized phrase.

Many educators believe that American students are too often taught to memorize and then repeat the information on a test. In the memorize-and-repeat lesson format, the teacher presents a concept or explains a procedure, and the student restates the concept or practices the procedure until they can repeat it on their own. O'Brien (1999) named this kind of instruction parrot math (and by extension, parrot science).

Memorization will not help our children become informed, productive citizens in the twenty-first century. With information increasing at ever-compounding rates, children need to develop knowledge that can be applied to a wide variety of new situations (U.S. Department of Education 2000). Instead, greater emphasis should be spent on children understanding math and science.

Shortcomings of Memorize-and-Repeat Instruction

The memorize-and-repeat view of learning has failed to get us where we want to go in math and science education. This teaching method is widespread in mathematics and science instruction (U.S. Department of Education 2000), and its results are alarming. Most American students, including the brightest, are not learning the most important ideas in math and science (Schmidt, McKnight, and Raizen 1997). Of the students from 41 nations taking part in the Third International Mathematics and Science Study (TIMSS), for example, U.S. eighth and twelfth grade students scored towards the bottom in general knowledge of science and mathematics (Stevenson 1998).

One way that traditional instruction leads to poor overall achievement is that children taught through these traditional methods have been found to memorize procedures and facts rather than construct an understanding of what they are doing (Wilson and Blank 1999; Gallagher

2000). This lack of conceptual understanding results in especially poor performance in solving the nonroutine problems typically found outside of school math and science (Dossey et al. 1988; Mitchell et al. 1999). Additional studies detail the failure of traditional teaching methods to promote problem-solving and reasoning abilities (Shaw and Jakubowski 1991; Cohen 1984).

ALTERNATIVES TO THE MEMORIZE-AND-REPEAT METHOD

Because memorize-and-repeat teaching methods do not prepare our children to face the as-yet-unknown problems likely in their lifetimes, reform in mathematics and science instruction is essential. In response to this need, educators have developed and tested teaching techniques that draw on constructivist elements. Four of these elements—the same elements that I found to benefit both my students and myself—have been the subject of extensive research:

1. The use of hands-on instruction, in which children are directly observing and manipulating their environment.

2. Giving children choice in what to learn and how to learn it.

3. Helping children reflect on their knowledge and efforts.

4. Teaching through problem solving and inquiry, in which students learn to solve a personally relevant problem or satisfy a curiosity.

What evidence supports the use of these elements in the math and science classrooms?

Hands-on Activity

Children learn math and science best when they are directly observing and manipulating their environment in a hands-on manner.

Children love to touch. For some, it is almost an uncontrollable impulse to reach out and handle objects in the world around them. Teachers well know the familiar sensations of students tugging at their sleeves, grabbing materials, or pulling at their shoelaces as they listen to a story on the floor.

In the era just prior to 1957, math and science instruction didn't allow for a lot of touching. Lessons often consisted of lectures, reading textbooks, and paper-and-pencil exercises. In response to the launch of the Soviet capsule *Sputnik*, American educational leaders pushed the use of more active forms of learning. Instruction that encourages *doing*—touching, moving around, manipulating, child-controlled movements, and observations of real objects—is supported by a tremendous amount of research.

Hands-on Instruction and Learning   Researchers have long examined the link between children's need to interact with their environment and of their learning of math and science. The mathematics educational research in the 1970s and 1980s focused on finding out the effects of hands-on instruction using manipulatives. Research known as meta-analysis, which combines many smaller studies to look for similarities, has been used to analyze the results of using math manipulatives in tens of thousands of children. The results are clear: "(manipulatives) have a high probability of increasing achievement, and appear to be essential in providing a firm foundation for developing mathematical ideas" (Suydam 1985 1). When compared to lessons where manipulatives are not used, lessons with manipulatives make it more likely that children will learn more, and this success applies to children of all ability and socio-economic levels (Sowell 1989; Suydam and Higgins 1977).

Research in individual classrooms has generally supported these findings. For example, one study reported on four third-grade classes, two taught using worksheets and two using manipulatives. The group using manipulatives significantly outscored the other group on a nationally normed test of conceptual understanding and computational skills, rising two grade levels above the worksheet group (Cain-Caston 1996).

Similar meta-analysis has been performed on science programs used in the 1960s and 1970s that stressed experimentation and the use of science manipulatives and tools rather than textbooks and lectures. When compared with traditional teaching methods, these activity-based programs resulted in significantly greater academic achievement and affect in science (Shymansky, Hedges, and Woodworth 1990). For example, in one meta-analysis involving 13,000 students in more than 900 classrooms, results showed an average of a 14% improvement in general achievement on standardized tests for students in the activity-based programs over nonactivity programs (Bredderman 1983). Other large-scale research found that students who engaged in hands-on activities every day or once a week outperformed students on standardized tests who had less time working hands-on (Stohr-Hunt 1996).

Again, smaller studies have supported this metaresearch. In a study of science instruction in a single classroom, third graders' understanding of air was examined after one of two types of lessons. The traditional group's lesson consisted of listening to a lecture, reading books, and filling out worksheets. The other group conducted 10 hands-on experiments. The students in the hands-on group outperformed the traditional group at lower- and higher-order learning. This success extended to children of varying ability levels (McDavitt 1994).

Outdoor field experiences have been a rich context for testing hands-on science. Research generally concludes that children benefit by the direct contact with the environment provided by outdoor field activities. In one such study, the use of the schoolyard as an outdoor classroom for studying environmental science was examined (Cronin-Jones 2000). One hundred ninety-one third- and fourth-grade students were assigned to either an indoor group or an outdoor group. Instruction for both groups included similar techniques: guided student reading, lecture, demonstrations, discussions, role-playing, and lab activities. However, the children taught indoors watched slide and film presentations and conducted their labs indoors; the outdoor children did their labs outdoors and substituted field observations of schoolyard plants and animals for the slides and films. After the 3-week unit, the outdoor group learned significantly more than the indoor group, based on a test of knowledge of plant and animal ecology. The researcher concluded that outdoor activities in which students are directly involved with their environment help children learn more than do media that are less hands-on, such as videotapes, books, and discussion.

The Standards and Hands-On Activity   The two primary national standards that address teaching, *Principles and Standards for School Mathematics* (National Council of Teachers of Mathematics 2000), and the *National Science Education Standards* (National Resource Council 1996) support the use of hands-on activity.

Principles and Standards outlines the advantages of hands-on instruction: "Representing numbers with various physical materials should be a major part of mathematics instruction in the elementary school grades" (p. 32). Specifically, "(c)oncrete models can help students represent numbers and develop number sense; they can also help bring meaning to students' use of written symbols and can be useful in building place value concepts" (p. 80). "But," the document cautions, "using materials, especially in a rote manner, does not ensure understanding" (p. 80).

Hands-on science is a major theme of the *National Science Education Standards.* "Teachers focus inquiry predominantly on real phenomena, in classrooms, outdoors, or in laboratory settings . . ." (National Research Council 1996, 5). This emphasis on real phenomena is echoed in each of the science content areas. For example, the life science content standard explanation begins, "During the elementary grades, children build understanding of biological concepts through direct experience with living things, their life cycles, and their habitats" (pp. 5–6).

State of Hands-On Instruction   Despite overwhelming evidence chronicling its benefits for children, numerous investigators have found that hands-on math and science is far from the norm in American elementary classrooms. A national sample of K–12 teachers, for example, found that only 73% of teachers reported using manipulatives at all, and only 62% used hands-on activity at least once a week (Henke, Chen, and Goldman 1999).

The science classroom is also an underused arena for hands-on activities. Teachers were found to use manipulatives in less than one of five science class periods in one study (Shaw and Hatfield 1996). Statistics from the 1996 National Assessment of Educational Progress (NAEP) found that only 56% of fourth-grade teachers used hands-on science activities as much as once or twice a week (O'Sullivan and Weiss 1999).

Even when teachers use hands-on activities, research shows that they are often practiced in ways that limit their benefit. The mere presence of manipulatives doesn't promote understanding (Cohen 1984), and teachers need to use them wisely for students to benefit (Sowell 1989). For example, if students use math manipulatives only in a rote manner, it can result in students using cumbersome calculation techniques rather than promoting understanding (Wheatley 1992). Other studies have found that the type of manipulative also matters, with some manipulatives proving more effective than others in helping children learn (Hiebert, Wearne, and Taber 1991).

Choice

Children benefit from being given the choice of what and how to learn.

Children demand choice. My daughter will offer, "Do you want to play Candyland or read a book?" "Ah, finally some choice for Daddy!" I think. But when I tell her I would rather read a book this time, she responds, "OK, but let's play Candyland!"

Math and science instruction in which students make frequent choices looks quite different than the more traditional teacher-directed instruction. Students can have choice in what topics to study, what questions to answer, how to solve problems, and how to present and interpret data. Research on choice centers on two areas: first, the effects of choice on motivation and, second, choice as a mechanism for differentiating instruction.

The Effects of Choice on Motivation
Contemporary motivation theory defines two types of motivation: *learning orientation* and *performance orientation*. Students motivated toward a learning orientation learn because they are curious about the answer. These students want to master a body of knowledge with understanding. In contrast, students driven by performance orientation learn in order to avoid appearing incapable or to get approval from others (Meece 1991).

Research suggests that how teachers teach greatly influence their students' motivation orientations. When teachers encourage student autonomy through, among other things, choice, they move students toward a learning orientation. Conversely, denying students this autonomy encourages a performance orientation (see Stipek et al. (1998) for a review).

The Effects of Motivation Orientation on Learning

Although a student motivated to learn, understand, and master a subject sounds like a good thing, does a learning orientation actually lead to increased learning? Intuitively, it seems that children will be more interested and work longer on activities of their own choosing. Studies on motivation confirm that students' learning orientation is accompanied by certain behaviors that potentially increase understanding. Students given instructional choices are more attentive and persistent, showing greater engagement in activities. They choose more challenging tasks and select more effective methods to solve problems (see Stipek (1996) for a review). Other research shows that providing students with a sense of autonomy through choice results in greater enjoyment of mathematical tasks, which in turn leads to greater persistence, flexibility, and creative behaviors (see Stipek et al. (1998) for a review).

In part due to these characteristics, students motivated in a learning orientation learn more than their performance-oriented peers (Ryan, Hicks, and Midgely 1997). Researchers have found that a learning orientation leads to greater conceptual understanding in mathematics but not necessarily to increased rote learning (Benware and Deci 1984). Other researchers (Stipek et al. 1998) have found the opposite to be true.

Choice as a Mechanism for Differentiating Instruction

Research indicates that when students are allowed to choose what to do and how to do it, they tend to choose things that make sense to them. Procedures and materials that don't seem helpful in solving a problem can be discarded in place of procedures and materials that do help. When children can choose levels of challenge and complexity that match their abilities, instruction is naturally differentiated to match each student.

One fertile area for research in the role of choice in promoting knowledge construction has been the results of encouraging children to develop their own procedures to solve problems, particularly procedures for computing. The traditional approach to teaching computation has been the teacher (or parent) modeling a method, and then the child practicing it until he or she has memorized how to do those problems. For example, teachers and parents often teach children how to add and subtract two-digit numbers the same way that they learned: In 67 + 34, first add the 7 and 4, put the 1 down there, carry the other 1 up here, then add all of these in the left column. Six plus three plus one equals 10, so the answer is 101. These standard algorithms are in wide use, both in school and among adults.

Alternatively, children can invent their own computing methods. Why let students choose and develop their own algorithms when standard algorithms are so efficient? According to constructivist theory (Kamii and Dominick 1998), standard, adult-developed algorithms may not make sense to children. If algorithms don't seem logical to the child, they may be perceived as just more meaningless procedures to memorize. Also, standard algorithms discourage logical thinking. Rather than wrestle with numbers in a meaningful way, children simply apply a ready-made procedure.

Research has revealed several advantages to having children invent their own computational strategies. First, children who use invented strategies are able to do more complex computations better than children taught traditional algorithms (Carpenter et al. 1998). This may be due to the conceptual understanding developed as children work to make sense of computation. Second, children who invent their own algorithms are also able to solve problems that are out of the ordinary more readily than children who just memorized the algorithms (Hiebert and Wearne 1996). It appears that constructed knowledge travels well, being able to be applied in a variety of situations. Third, students develop a sense of number, place value, and operations as they experiment with and debate invented algorithms (Carpenter et al. 1998). Fourth, students' thinking becomes more flexible as they are exposed to a variety of invented procedures (Carroll and Porter 1997). It has even been found that children taught standard algorithms lose some of the conceptual understanding that they had prior to instruction (Narrode, Board, and Davenport 1993; Kamii and Dominick 1998).

Clearly, children allowed to choose computational strategies that make sense to them develop greater understanding. But can they solve straightforward addition and subtraction problems? Intuitively, it would seem that more traditional memorize-and-repeat, drill-and-skill methods lead to more accurate computation of standard problems. However, a formidable body of research has concluded that children encouraged to invent their own procedures either beat or match children taught through more traditional means in computation (see Kamii and Dominick 1998 for a review).

Research on the role of choice in science instruction has generally focused on the choice imbedded in inquiry. These data are presented later in this chapter. Several studies have examined the role of choice outside of inquiry. For example, one concluded that reasoning ability among second graders was increased when students were encouraged to follow their own interests while manipulating science manipulatives. Those students who strictly followed the teacher's direction while working with their objects developed less complete classification understandings than those students who could choose their actions (Cohen 1984).

The Standards and Choice   Choice is a common theme in both national standards. The math standards state the importance of giving choice in how to solve problems: "Classrooms in which students have ready access to materials such as counters, calculators, and computers and in which they are encouraged to use a wide variety of strategies, support thinking that results in multiple levels of understanding" (National Council of Teachers of Mathematics 2000, 119).

The science standards describe successful science classrooms as ones where "teachers and students collaborate in the pursuit of ideas, and students quite often initiate new activities related to an inquiry. Students formulate questions and devise ways to answer them, they collect data, and decide how to represent it, they organize data to generate knowledge, and test the reliability of their knowledge they have generated" (National Research Council 1996, 7). The role of choice in motivating students and teaching responsibility is also noted: "Teachers (should) give students the opportunity to participate in setting goals, planning activities, assessing work, and designing the environment" (p.15).

Wise Choices   When students are allowed to choose, how can teachers be sure that these are good choices? Some educators report that when students choose their own activities, "they find their own thinking levels" (Phillips and Phillips 1994, 52). Children, like adults, are drawn to what seems logical or needs a bit of puzzling to figure out. Activities that are far too easy are boring, and those so far out of reach of children are frustrating. Therefore, children can be counted on to choose developmentally appropriate and interesting activities. Other researchers (Ames and Archer 1988) have also concluded that, in a positive, supportive classroom environment, choice and other strategies that give children autonomy tend to make children more willing to take risks and challenge themselves. The concept that choice leads to selection of activities at a reasonable level of challenge is supported by research showing that when students feel competent and in control—as when they are allowed to choose—they enjoy those activities more often (Deci and Ryan 1985).

It's apparent, however, that situations exist where not every child will choose challenging, appropriate activities. In one study (Meyer, Turner, and Spenser 1997), researchers followed two groups of students during a math unit: *challenge seekers* and *challenge avoiders*. The challenge seekers reported a tolerance for failure and tended to be more learning oriented. The challenge avoiders did so out of fear of failure and were more performance oriented. Giving students choices can be a powerful instructional tool, but teachers need to monitor students' choices to ensure that students make choices that will benefit their learning.

State of Choice   Little research has been published that explores only the extent of choice in the classroom. However, it can be construed from research presented earlier on the prevalence of memorize-and-repeat instruction that choice is not widespread. It appears that children are getting more opportunity to practice what their teacher tells them rather than choosing their own way of doing things.

A second indicator that there is limited choice in U.S. classrooms comes from research on the state of problem solving and inquiry. Because both processes call for a certain degree of choice, students who frequently solve problems or engage in inquiry are often required to make choices. Unfortunately, problem solving and inquiry are far from universal in the math and science classroom, as illustrated in the upcoming "State of Problem Solving and Inquiry" section.

Reflection

Children benefit from reflecting on their knowledge and efforts.

Having time for reflection is a shrinking commodity for children. Students in many classrooms are rushed from activity to activity, with little time to think about what they have done. It is little wonder that Nothing is the stock response to What did you do at school today?

Reflection can be defined as examining one's own thoughts and actions. Although it is often seen as a solitary activity to be done in one's head, reflection is facilitated through writing and talking. Both encourage the mental processing of experiences, turning nebulous thoughts concrete as students organize and connect experiences to past learning (Zemelman, Daniels, and Hyde 1998; Borasi and Rose 1989). The use of metacognitive journals, learning logs, and discussion are tools often used by teachers who encourage student reflection.

Writing about Math   The effect of reflective writing about math, in particular, has been explored. Research suggests that as students write about math, the interaction between language and thought helps develop understanding. The potential advantages of reflecting about math through journal writing include increased affect, greater knowledge of mathematical content, an improvement in learning and problem-solving skills, and a more accurate view of the nature of mathematics (Borasi and Rose 1989).

There is evidence to support at least some of this potential. In one study (Jurdak and Abu Zein 1998) 104 middle school students responded in math journals to open-ended prompts for 7 to 10 minutes each lesson. Prompts were provided that were either content-oriented (Give your friend a summary of the important parts of today's lesson.), or affect-oriented (How do you feel you did on the last test?). After 12 weeks, these students developed greater conceptual understanding, procedural knowledge, and mathematical communication than students who spent that time doing additional exercises.

First-graders who wrote in math journals connected new mathematical experiences to prior experiences, suggesting that writing can aid in constructing new knowledge (Wason-Ellam 1987). Other researchers (Davison and Pearce 1990) have reported that writing appears to increase general mathematical achievement.

Discussing Math   Reflection through discussion can also help lead students to mathematical understanding. As students express their beliefs to their peers, they are often called upon to clarify and defend their ideas. This leads students to amend and refine their thoughts (Ball 1993). Research suggests that the verbal reflection that occurs as students share insights and strategies results in better reasoning and computational skills (Carroll and Porter 1997; Baroody 1999).

Studies examining the role of class discussion have found that this expression leads to better mathematical problem solving. One study (Hiebert and Wearne 1993) looked at the effects of problem solving and class discussion on six second-grade classes, which were divided into two groups. Instruction in the first group was primarily written-symbol, paper-and-pencil problem solving, and short-answer, teacher-led discussion that often called for students to merely recall facts and processes. The second group also solved problems, but most were story problems that encouraged deeper discussion, in which students explained and defended their problem-solving strategies. Students involved in the deeper discussion developed significantly better understandings of addition, subtraction, and place value.

As children and teachers collectively reflect on and communicate about their math experiences, they can negotiate conclusions that are both acceptable to the group and accurate. For example, one study followed second-grade students who were taught math through an instructional model in which students discussed and justified their solutions to problems (Cobb et al. 1991). At the end of the year, the project students developed greater conceptual understanding than a control group that received more traditional instruction. In a follow-up study, the project second graders were tested a year later, after attending traditional third-grade classrooms. They maintained their advantage over their nonproject peers in conceptual understanding (Cobb et al. 1992).

Writing and Discussing Science   The role of reflection has also been explored in science education. Researchers have found that hands-on experiences by themselves don't necessarily lead to greater understanding of science (Schauble, Klopfer, and Raghaven 1991; Cohen 1984). Mechanically pushing children through hands-on activities without opportunities to process the experiences has been referred to as "activitymania" (Moscovici and Nelson 1998). It has been suggested that reflection may help bridge the gap between concrete experiences and understanding. As children reflect, they can link new experiences with past knowledge to create new understanding.

Reflection through writing has been promoted by educators. Two models of science writing—the knowledge-telling and knowledge-transforming models—have been articulated (Bereiter and Scardamalia 1987). The knowledge-telling model, where the writers organize and write down as much as they remember about a science topic, occurs as children write test essays and describe observations. Although knowledge-telling writing can be used as an assessment and for reporting observations, the authors speculate that it does little to generate new knowledge. In contrast, knowledge-transforming writing, such as interpreting the results of an experiment or speculating on the reasons behind an observation, is more likely to promote new understanding as students connect prior content knowledge to new experiences. When the writing is done for an audience, new understanding may especially be fostered, because the writer must consider the language and syntax to best communicate the content (Keys 1999).

Research documents some of these benefits of writing and discussion. Writing that requires students to explain the results of hands-on activities increases understanding of the scientific concepts being demonstrated (Palincsar, Anderson, and Daniel 1993). Other studies imply that students learn more through the discussion generated as students cooperatively solve problems and work to incorporate new phenomena into existing science beliefs (Glasson and Lakik 1993; Resnick et al. 1991).

Some research has focused on the use of both discussion and writing. In a study of a fifth-grade class, students discussed and wrote about science (Peasley, Rosean, and Roth 1992). The researcher concluded that when discussion and writing took place within a backdrop of genuine inquiry and student-directed learning, students developed a deep conceptual understanding that they were able to apply in a variety of situations.

The relative benefits and interplay of talk and writing have also been the topic of research (Rivard and Straw 2000). In this study, middle school students were involved in an ecology unit in which hands-on activities, library research, worksheets, and other teaching methods were used. In addition to this instruction, problem-solving activities in which students had to apply their knowledge to new situations were given. A talk-only group discussed the problems with their peers but did no writing. A write-only group responded in writing to the problems without prior discussion. A talk-and-write group discussed the problems as a group and then individually wrote their responses. The results suggested that although peer discussion may increase simple recall of facts, combining talk with writing facilitates understanding of more complex concepts. Writing alone was not found to be as beneficial as talk alone. The researchers suggested that student talk increases the amount of information available, which can then be processed through writing.

The Standards and Reflection   Both national standards stress the importance of reflection. In math, the role of language in building mathematical knowledge "is a very powerful tool and should be used to foster the learning of mathematics. Communicating about mathematical ideas is a way for students to articulate, clarify, organize, and consolidate their thinking." (National Council of Teachers of Mathematics 2000, 127). Reflection also encourages student thinking habits: "As teachers maintain an environment in which the development of understanding is consistently monitored through reflection, students are more likely to learn to take responsibility for reflecting on their work and make adjustments necessary when solving problems" (p. 54).

The science standards stress that, by its very nature, science "often is a collaborative endeavor, and all science depends on the ultimate sharing and debating of ideas" (p. 5). The standards also underscore the importance of reflection: "An important stage of inquiry and of student science learning is the oral and written discourse that focuses the attention of students on how they know what they know and how their knowledge connects to larger ideas, other domains, and the world beyond the classroom" (p. 7).

State of Reflection in School   Despite the evidence showing its benefits, relatively few teachers encourage substantial reflection in their classrooms. In cross-cultural studies of mathematics instruction, researchers have found numerous unrealized opportunities for reflection in American classrooms (Stigler and Perry 1990; Yang and Cobb 1995). A survey of fourth- and eighth-grade teachers found teachers place greater emphasis on solving routine problems than on discussing mathematical ideas (Mitchell et al. 1999). Reflective writing is also underused. One survey of 75 elementary school classrooms found that less than 10% of writing was content area writing, including math and science (Sunflower and Crawford 1985). In a more recent survey of 117 teachers, 63% either had never heard of writing in math class, or used it rarely to never. Ironically, all of the surveyed teachers were members of the National Council of Teachers of Mathematics (Silver 1999).

Considering the lack of reflection in many classrooms, it is no surprise that American students struggle with mathematical communication. Analysis of both the NAEP and the TIMSS mathematics tests (Mitchell et al. 1999; Wilson and Blank 1999) determined that those areas involving written and drawn reasoning and communication proved among the most challenging areas for students. These same researchers concluded: "Communication skills should be central to the activities in every mathematics classroom, and not simply relegated to the ubiquitous direction of 'show your work.' " (Wilson and Blank 1999, 47). The notion that reflection needs to be substantial is supported by research that found no correlation between students' success at word problems and the frequency in which they wrote a few sentences on how to solve a math problem (Mitchell et al. 2000).

Problem Solving and Inquiry

Children build math and science understanding best as they solve personally relevant problems and through inquiry into what interests them.

Problem solving and inquiry are at the heart of learning. Humans constantly face problems that are important enough that we want to solve them. The oil-polluted schoolyard ditch that was related earlier is an example of a problem that engaged me. At other times, there is no problem, just a situation that activates our curiosity. Curiosity about a stick lying at the side of a stream—Will it float?—is satisfied through inquiry. The child designs an investigation, such as throwing the stick into the water, which leads to an answer. Often the results are shared—Look, it's floating under the bridge!—and new questions are generated: I wonder if we can sink it with some rocks?

Teachers who use problem solving and inquiry allow children choice. In problem solving, the problem is usually given, but the solution method is not. Inquiry requires children to choose a question, design an investigation to answer that question, and to present and debate the findings.

Mathematical Problem Solving   Children solving problems is both a means and an end to math education. Studies examining the effects of instruction emphasizing problem solving compared to more traditional instruction have concluded what may seem to be obvious: problem-solving classrooms produce students better able to solve problems (Carpenter et al. 1989; Cobb et al. 1991). For example, a districtwide program designed to help lower-performing students solve story problems resulted in a significant decline in the percentage of children who scored in the bottom fifteenth percentile on a standardized math test. Instruction focused on solving daily, student-generated story problems and careful analysis of problem structure (Butler 1991). Another study concluded that emphasis on solving problems and understanding principles seemed to increase math achievement among middle school students (Hoffer and Gamoran 1993).

The notion that the problem-solving process builds more than just the ability to solve problems is also supported by research. In one study of forty first-grade classrooms, students of teachers who used problem solving more than direct instruction on number facts and who encouraged children to use a variety of problem-solving strategies outperformed control students in knowledge of number facts (Carpenter et al. 1989). The effects of open-ended problem solving along with an emphasis on understanding concepts rather than practicing procedures were the focus of another study (Cai, Moyer, and Grochowski 1997). Two sixth-grade classes studying averaging were taught in this way. Students increased both their ability to accurately compute averages, and to explain their use of a number of strategies. The students were also able to apply their knowledge to solve novel problems. Problem solving also has an effect on children's mathematical beliefs and attitudes. Children in problem solving classrooms believe that there is more to math than getting the right answer quickly. They value the development and communication of a range of strategies. They take responsibility for their own learning, evaluating the accuracy of their own answers instead of looking to the teacher as the sole source of feedback (Franke and Carey 1997)

Inquiry Science   Research on inquiry-based science indicates numerous advantages over other forms of teaching. First, students develop greater understanding of science concepts and processes when teachers use inquiry-based practices (Chang and Mao 1999; also see Haury (1993) for a review). One meta-analysis of 140 studies of middle and secondary school science instruction examined the effect of inquiry-related practices on student achievement (Wise 1996). The author looked at the effects of student questioning, use of manipulatives, and strategies that focus students on the purpose of instruction—strategies that are associated with inquiry-driven instruction. Students taught through these instructional strategies performed an average of thirteen percentile points higher than students not taught through inquiry-driven methods.

Another study compared sixth- and seventh-grade students taught through a laboratory-centered inquiry program and those taught through traditional textbook methods. The inquiry group developed greater laboratory skills, science process skills, and general knowledge and understanding of science (Mattheis and Nakayama 1988).

A second advantage of inquiry is that it results in greater affect than traditional instruction. Not surprisingly, researchers have found that students who work to answer their own questions and share their knowledge with their classmates feel positive about science. For example, middle school students taught earth science through inquiry-based methods had a better attitude toward learning science than those taught through teacher-directed lectures, textbooks, and demonstrations (Chang and Mao 1999). Other research has also found greater affect during science inquiry (Rakow 1986).

A third advantage of inquiry science is that it appears to reach groups typically underserved by traditional science instruction. In a study of 172 fourth-graders that included a large group of learning disabled students, science instruction was through one of two hands-on programs—one inquiry-based and one not inquiry-based. Those taught through inquiry—children both with and without learning disabilities—understood science content better than those students in the non-inquiry group (Dalton et al. 1997).

Researchers examining the effects of problem-based learning in science for high-ability students have found enhanced motivation among those students (Van Tassel-Baska et al. 1998) and increased ability for finding problems (Gallagher, Stiepen, and Rosenthal 1992). Programs that encouraged original student investigations, another hallmark of inquiry teaching, were found to be effective with high-ability students (see Van Tassel-Baska et al. (1998) for a review).

American Indians are another group whose needs are typically not met as instruction moves away from real-life contexts (Leap 1982). These students may benefit from the personally and culturally relevant instruction often found in inquiry science (Cajete 1986) and math (Mather 1997).

Inquiry-based instruction has also been found to be beneficial for English as a second language students (Rosebery, Warren, and Conant 1992), and deaf children (Boyd and George 1971). In the special education classrooms, students taught through inquiry learn more, remember it longer, and prefer inquiry to textbook methods (Scruggs et al. 1993). Other research suggests that an inquiry science can help encourage students other than white males from participating in science (Kahle and Damnjanovic 1994).

The Standards and Problem Solving and Inquiry   Problem solving as an instructional element is widely supported in the national math standards. In the process of problem solving, "students must draw on their knowledge, and through this process, they will often develop new mathematical understandings. Solving problems is not only a goal of learning mathematics but also a major means of doing so. Students should have frequent opportunities to formulate, grapple with, and solve complex problems that require a significant amount of effort and should then be encouraged to reflect on their thinking." (National Councel of Teachers of Mathematics 2000, 51).

The science standards are very explicit about the importance of inquiry as an instructional element: "Inquiry into authentic questions generated from student experiences is the central strategy for teaching science" (National Research Council 1996, 5).

State of Problem Solving and Inquiry in Schools   The research documents the many benefits of students working to satisfy curiosities and to solve problems. Unfortunately, it appears that this evidence has gone unheeded in most schools. Part of the TIMSS project was to videotape a random sampling of eighty-one U.S. eighth-grade teachers as they taught a math lesson. Even though much of the time students in these classrooms were working on problems, problem solving as described in the standards—"engaging in a task for which the solution method is not known in advance" (National Council of Teachers of Mathematics 2000, 52)—rarely took place. Instead, students spent almost all their time memorizing and repeating; that is, practicing the procedure that the teacher had demonstrated (Stigler and Hiebert 1997). During science, reading and cookie-cutter experiments are often used in place of inquiry (Lockwood 1992a, 1992b), and results of the TIMSS suggests that more experimental science activities are needed (U.S. Department of Education 1996).

Even when they are used, the effectiveness of problem solving and inquiry is greatly influenced by how teachers use them. For example, research shows that solving fewer problems in depth is better than superficially touching on many problems (Hiebert and Wearne 1993). Similarly, science inquiry can be conducted using primarily secondary sources—books, videos, and experts. Although these sources offer excellent opportunities to extend knowledge, research suggests that "students are likely to begin to understand the natural world when they work directly with natural phenomena, using their senses to observe and using instruments to extend the power of their senses" (National Science Board 1991, 27).

CONCLUSION

Three major conclusions can be drawn from the research on the teaching and learning of mathematics and science. First, too many U.S. children are leaving school with insufficient math and science knowledge, and this is primarily due to inadequate preparation in school. Second, children are much more likely to understand these subjects while involved in hands-on activities, where choice and reflection are supported and where instruction revolves around problem solving and inquiry.

With this evidence in hand, a solution to America's math and science illiteracy seems as easy as 2 + 2; we know how children learn these subjects, so let's teach that way. A third conclusion from this research, however, reveals a paradox: the very elements that appear to help kids learn better are not used in most K–2 classrooms across the country. The next chapter suggests some possible reasons behind this paradox and shows how teachers can overcome it.

3

Connecting Math and Science Teaching to Literacy Instruction

For 2 years straight, as a bachelor right out of college, I had either take-out fried rice or frozen pizza every night for dinner. I consistently ate these high-fat foods despite the fact that I was health-conscious and did not particularly enjoy either food that much. This illustrates—much to my arteries' chagrin—that change can be very, very difficult. And if food choices are resistant to change, consider how hard it is to change what you do in the classroom. Teaching techniques are practiced to perfection and then become habit. This may be why typical math and science instruction has not changed significantly since you were 7 years old.

The teaching described in this book is quite different than the norm. You may find that these ideas fit easily into your repertoire. Or, you may have to make considerable changes to your math and science teaching. The good news is that reform math and science teaching is not too much different than what you may already be doing so well as a teacher of reading and writing. In other words, these changes can be made easier by linking it to something with which *you are already successful*.

This chapter details the parallels between reform math and science teaching and reform reading and writing practices (see Zemelman, Daniels, and Hyde (1998)) that are becoming widespread in American classrooms. Specifically, learning how to use math and science workshops will be easier if you already teach reading or writing using workshops. You are likely to be more comfortable and effective teaching math and science when you can apply what you already know. Similarly, what works in reform math also works in reform science; the teaching of both subjects can be transformed simultaneously.

REWORKING THE WORKSHOP

What are those good things that we do during literacy instruction that can inform our math and science teaching?

Children Learn by Doing

In school I learned to write by diagramming sentences. In fact, we diagrammed so many sentences, turned adjectives into adverbs by adding *-ly*, and distinguished between metaphors and similes, that we had very little time left over for *writing!* This may explain why I got to college not knowing how to write much beyond letters home.

Fortunately, things have changed; we have come to know that children learn to write by writing. They do not have to be drilled in every skill before they can put pencil to paper. Instead, children learn the basics by thinking and writing about ideas important to them, under the guidance of a knowledgeable teacher.

Unfortunately, the idea of learning by doing has not yet filtered into every math and science classroom. Children are often drilled in skills outside of the larger context of math and science. For example, children are often taught how to add numbers before they get a chance to solve story problems requiring addition. What we need to keep in mind is that children have had years of experience solving real-life addition problems even before they enter school. Then they are told by well-meaning teachers, You don't know how to solve story problems yet, but don't worry, I'll teach you how to do it! In the process, children come to see addition and other such skills as divorced from any real context.

The research cited in Chapter 2 shows that skills are taught better through engaging in the larger context of math and science—solving interesting problems and satisfying curiosities through inquiry. When children need to add to determine who won a math game or to see how many guppies were born overnight, they see addition as a useful tool. They are more motivated to learn and more likely to use their own understanding as a base on which to construct new addition knowledge.

Memorization Has Major Limitations as a Learning Tool

The number of words that children see in print during the primary grades likely reaches into the thousands. It would be impossible for children to memorize the spellings of all of those words, so how do we ensure that our students learn how to read? We have children look for patterns in words, encourage them to get cues from the contexts of the sentences, and teach children that they can look for words that they do know hidden inside unfamiliar words. And we have children read, read, read; many words are learned from long-term exposure to books that kids like.

Just as we would never send home a list of a hundred words for children to work on every night, neither should memorization be a keystone of our math and science instruction. When more time in the classroom is given to the definitions of *photosynthesis* or *triangular prism* than to understanding these concepts and children spend hours in class and at home memorizing the addition and subtraction facts, it is easy to see why some children (and adults) see math and science as just a string of unrelated things to memorize and then repeat.

To develop knowledge that is useful for more than a game of Jeopardy!, the same practices that we use to teach reading can be applied to math and science. Addition and subtraction facts, for example, do not have to be just memorized. Facts can be solved through patterns (any number plus one is always the next number), through the context of the problem (There are 4 fish here and 3 over there . . . that's 5, 6, 7 . . . 7!), and through using familiar facts to get unfamiliar ones (I know that 6 + 6 is 12, so 6 + 7 must be 13). And just as children learn to read by reading interesting books, the more children engage in solving intriguing problems, the more familiar they will be with the basic math facts. Children who learn the facts this way tend to be able to recall them automatically *with understanding*. Also, the whole process of thinking rather than memorizing offers opportunities to strengthen number sense and knowledge of place value.

Children's Inventions Spring from Their Understanding

Mi dud is the garatis! We know that this kindergartner's message to his father is good writing, and we encourage invented spelling. We know that children's invented spellings come from their understandings of letter sounds. For every child at some point 33,333, gt, garatis, and gratest are all important steps toward greatest.

Young children can also invent their own procedures for solving problems and making inquiries in math and science. Like invented spelling, children's invented procedures often are different than adult ways of doing things. For example, a child's invention of a way to visually represent quantity is shown in Figure 3–1. Although it may not conform to conventional, adult graphing, it is based on that child's understanding and is excellent work for that child at that time in her development. Research indicates that many young children have not yet developed the spatial abilities needed to understand standard graphs (see Friel, Curcio, and Bright (2001) for a review), so encouraging these inventions is preferable to insisting that children use "correct" graphing technique.

This is not to say that children's inventions should always replace standard procedures. Like invented spellings, invented procedures are a step in the right direction. Teachers then need to support children's development toward accepted, efficient math and science.

Invented spellings are technically incorrect, but math and science procedures that children invent often yield correct results. Children have

Do you like choclate or caramel better?

Caramel

Chocolate

FIGURE 3–1   A child's invented graphing technique.

years of experience with the physical world before they spend much time with print. These early experiences sharing cookies and racing toy cars often give them the background for figuring out their own ways of dividing numbers and controlling variables. These invented procedures are to be encouraged because they are built on understanding rather than memorization.

A corollary to this is that it is usually better not to show children how to solve problems or run inquiries. It is tempting to teach children "a better way" to do something, especially when their invented strategies seem exasperatingly slow or inefficient. Keep in mind, however, that invented ways of doing things usually spring from understanding. Expect and nurture what children do naturally.

Children Should Draw Their Own Conclusions

After reading *The Giving Tree* (Silverstein 1987) to my first and second graders, I asked them what this story was all about. Some gave the literal meaning: it was about a tree that loves a boy and gives him things. Other students said that the book showed how people use trees in real life to make things. Still other children were sure that it was about how you should not take advantage of your friends. I held a slightly different view than the children, the author probably had other thoughts, and had you read the book, you very well may have come to altogether unique conclusions.

So, who is right? No one, and everyone. We know that children's experiences influence what they get from stories, and that this is to be encouraged. It shows that each child is using his or her abilities, taking ownership of the book, and not relying on the teacher's opinion which the child does not necessarily understand.

But even though we let children infer their own conclusions in reading, we too often insist that children come away with identical conclusions from math and science activities. For example, after exploring how magnets interact with different materials, we might be inclined to bring children to the whole group and tell them what they just learned. You may have noticed that magnets just stuck to some metal items. Magnets are attracted only to iron. Meanwhile, the children are scratching their head wondering, "Is *that* what I leaned? I thought I learned that I could push away Ambracia's magnet with my magnet. "Or I learned that my magnet could hold 10 paper clips. Whoops, I'd better try to remember what the teacher said. It must be the right answer." Clearly, conclusions that children draw from math and science activities will vary. And this is a good thing. It shows that children are not trying to memorize what the teacher said but instead are using their own unique backgrounds, interests, and developments to understand their world.

This point can be illustrated by doing the following activity with a friend. Light a candle and, without talking, write down everything that you observed about the burning candle. Then compare notes with your partner. The odds are that each of you will have come away from the activity with different conclusions. Maybe you focused on the smoke's movement, whereas your friend noticed that the candle sputtered with each slight breeze. Perhaps the flame's layers of color reminded you of a favorite painting, but a campfire experience influenced your friend. So who is right? Each of you came away with something important, legitimate, *but different* by looking at the exact same candle. So should it be with your students.

This analogy has its limitations. Although it is important to not press "the correct" answer on students, it is just as important to present and debate children's conclusions so that they can perhaps change ideas that are clearly incorrect. If a child concludes that 5 plus 6 equals 10, you might ask, "Well, what's 5 plus 5? You say that 5 plus 5 is 10, so can 5 plus 6 also be 10?"

Choice Is Essential

SSR (sustained silent reading) is a sacred time for most of my students. We read throughout the day, but SSR holds a special place among children because they get to choose. Nearly every book, space, and reading partner is fair game during SSR, and that gives kids the sense of freedom and control that they crave. It also helps ensure that children are picking books that match their needs and interests.

Gone are the days when teachers dictate every book that children read, and we must extend this same level of choice to our students during math and science. It is not uncommon to hear instruction like this: To add 45 and 22, first add the ones, and then the tens, or You are going to show how solids interact with liquids by putting raisins in water. This not only denies children that sense of control, but the teacher's thoughts on how to do things may not match the students' thoughts. Students then have to memorize rather than construct their own knowledge.

Better approaches allow kids to choose procedures and topics that fit each individual. We need to provide choice in the topics that individuals study, in the ways that they solve problems, and in the partners and materials with which they work. To add 45 and 22, children can develop their own algorithms; adding the ones first and then the tens is an adult-invented procedure that is counterintuitive to the way children learn. Similarly, children can be allowed to pursue their own ideas to explore how solids and liquids interact.

Kids often choose topics and procedures that do not always make sense to the teacher. Sometimes this is OK; like the advanced reader who chooses a too-easy book during SSR, a child's choices usually fulfill some need. At the same time, you have to be aware that some topics and procedures are not worth children's time. A skilled teacher learns to distinguish between worthwhile alternative topics and procedures and ones that are impractical or probably won't lead students anywhere.

Sharing Encourages Reflection

Most children love to share their writing. The sharing time at the end of writing workshop is always too short to accommodate all the children wanting to read about their trip to the amusement park or their own version of a Frog and Toad story. We know that student sharing gives confidence, exposes students to new ideas, and helps develop a community of writers.

We also need to build in regular and substantial amounts of time for children to share during math and science. This sharing time is not just to make kids feel good. Children need to practice communicating mathematically and scientifically. One child's ideas can stimulate thinking in both that child and others, and a community of mathematicians and scientists will develop as students present, question, and debate their work. And we should also recognize that student communication is essential to planning for and reflecting on our own teaching. Thoughts that children jot in a learning log or share in a discussion can point to what they are interested in, what misconceptions are out there, and the effect our lessons have on their learning.

BRINGING IT ALL TOGETHER

Teachers who want to incorporate more of these elements into their math and science instruction face two challenges: (1) It is hard to teach in a way that is very different than how we currently teach, and (2) parents and administrators may resist teaching methods that look so different from traditional teaching.

How can teachers overcome these obstacles? Fortunately, we are not trailblazing new ground here. We have already demonstrated that we can substantially change how we teach. As a rule, we do not teach reading and writing the same ways as we did 10 years ago; we have adopted instructional elements that help children read and write better. These elements were brought into the classroom through, among other ways, the workshop lesson format. In other words:

It is time to rework the workshop.

Literacy workshops were designed by educators (Graves 1983; Calkins 1986; Atwell 1987) to maximize the use of elements that many teachers intuitively knew—and research had confirmed—helped students learn better.

Reading and writing workshops include a minilesson, followed by an activity period, and concluding with a reflection period. During the minilesson, the teacher illustrates different aspects of the reading or writing process, conventions, or thinking strategies and then sets expectations for the activity period. For example, a writing workshop topic could focus on the use of language. The teacher might share passages from published or student writing that contains such language during the minilesson and perhaps have the class generate a list of descriptive words. Then the teacher may instruct the students to try to use some of these words in their writing.

The activity period is for students to work on writing activities while the teacher conferences with students about their writing. Using our sample workshop, students would choose among different writing formats—fiction, poetry, lists, etc.—and the teacher would meet with individual students, focusing on each child's needs.

The reflection period begins with students sharing their books and thoughts with the entire class. If descriptive language is the workshop focus, discussion might include how students were able to work the language into their writing.

Today, literacy techniques such as workshops are becoming common in American classrooms (Zemelman, Daniels, and Hyde 1998). Their favorable reception suggests both that teachers feel successful using them and that parents and administrators have accepted them.

Math and science can also be taught through workshops. Workshops offer several distinct advantages over more conventional teaching methods. First, the workshop format encourages the systematic use of hands-on activities, choice, reflection, and problem solving and inquiry—elements proven to promote understanding. Second, the workshop allows good teachers to extend what they do so well already—letting children learn by doing, making choices, drawing conclusions, sharing, and understanding rather than memorizing, all in a developmentally appropriate ways—into math and science. Third, you may already use the workshop for literacy and would feel more confident and comfortable teaching math and science workshops. Fourth, parents and administrators are more likely to accept the movement away from traditional math and science instruction, because the format and philosophy is similar to the literacy workshops already in use in their schools.

Most importantly, with math and science workshops, children can finally abandon "parrot math and science" as they are allowed consistent access to the keys for understanding these subjects. It is time for us to bring our best literacy teaching techniques, lessons learned from our personal experiences, and results of research into math and science instruction. Math and science workshops do just that.

4

Planning your Math and Science Programs

My wife and I enjoy camping. Let me rephrase that—we have been camping, and many of these trips rank among our greatest experiences. But we have also lived through trips that made us yearn for the relative pleasures of home, such as cleaning the gutters. For us, the difference between good and bad camping comes down to planning. We had to reserve the spot months in advance in order to pitch our tent along a riverbed in Rocky Mountain National Park. On another trip, we sweated out a night in our 90° tent, mentally replaying the moment when we discovered that there wasn't a shower within 10 miles. As sure as campfire omelets taste better if you don't forget to pack some eggs, thorough planning is key to great camping.

We all know that the same goes for teaching. A good math and science program begins and ends with plans. Planning the overall curriculum and arranging workshops within each unit of study is the focus of this chapter. How to plan individual workshops will be presented in Chapters 5 through 8. Chapter 9 addresses how reflecting on assessment results affects future plans.

PLANNING CURRICULUM

Curricular planning refers to the scope and sequence of units throughout the school year as well as to planning individual units of study. The division of math and science into units is somewhat arbitrary. My district's main math resource is a textbook consisting of about ten units for each grade level. Each grade also has three prepackaged science units. Although I do not teach straight from these resources, the scope and sequences are workable, and I follow them to some degree. I find it easier to use the available activities and materials as a starting point than to develop my own scope and sequence, which may be no better.

Curricular planning addresses a number of important issues, especially

1. What is important for children to learn?

2. How can they learn these important ideas?

What Is Important for Children to Learn?

What are the important topics in math and science that children need to learn? What ideas—concepts, procedures, facts, and attitudes—within topics are vital for each unit? Answers to these questions can come from many sources:

• National Standards: The national standards are probably the most comprehensive, research-based sources to guide the selection of unit topics and objectives. Committees of eminent educators and community leaders created them, with the purpose of preparing children for future personal, social, and economic challenges (Lee and Paik 2000). The three major standards are *Principles and Standards for School Mathematics* (National Council of Teachers of Mathematics 2000) for math, and for science *National Science Education Standards* (National Resource Council 1996) and *Benchmarks for Scientific Literacy* (Project 2061 1993).

• Personal: You, your teammates, your principal, the parents of your children, the local community—all have personal opinions about what should be taught in math and science. Although many of these ideas may be worthwhile, letting personal opinion alone dictate units and objectives is not likely to lead to a comprehensive, research-based curriculum.

• Program: Publishers of math and science programs select their own units and objectives. Program publishers generally put greater emphasis on making teachers feel comfortable than on adhering to the national standards—they need to sell lots of books, after all—and thus many are only tenuously linked to the national standards. Because marketability so often competes with what is right for children, it is important to look critically at what program publishers offer.

• School or District: "In many schools and districts, the local curriculum is a hodge-podge of individual initiatives knit together by collective good intentions" (Carr and Harris 2001, 1). Curriculum that is pieced together by various committees and individuals—often years earlier—is not likely to be aligned with current national standards.

• State: Each state has goals and standards. Often these align with the national standards and are very broad, which frees you to use the national standards and still fulfill state requirements.

Dealing with Competing Agendas   With the knowledge that the national standards are the preeminent guides, choosing objectives for a unit should be easy, right? Unfortunately, this is not usually true. Often these different sources conflict with each other, and you can find yourself torn between what you believe and what is required of you. Dealing with these competing agendas can be a tricky business. I am very fortunate to be in a progressive district, but even here conflicts arise. Our science objectives, for example, put equal emphasis on all three states of matter, even though both national standards make it clear that, based on research, young children are not likely to understand the concept of gas (Project 2061 1993).[1]

Resolving these conflicts takes creativity. For example, I had students read books about the different states of matter, including gas during our language arts time. While I helped children draw connections between the book and their science workshop experiences with solids and liquids, I also made students aware that there is another state of matter. I introduced some of the vocabulary associated with gases, and we discussed some situations in which children "see" gas, such as when blowing up a balloon. Using these nonfiction books in this way allowed me to teach important language arts ideas as children expanded upon their understanding of solids and liquids gained through hands-on inquiries. And I nominally covered the concept of gases.

Presenting your concerns to administrators may be another way to resolve competing agendas. Principals are often aware that district and state objectives can be out of step with the national standards and may encourage you to forge ahead until local objectives catch up. There is also power in numbers. Getting fellow teachers aboard to present concerns may help persuade administration to negotiate a compromise.

Large-Scale Versus Small-Scale Reform   Even if these competing agendas can be resolved, picking objectives is still a challenge. The sheer number and scope of objectives for the national standards indicate the need for carefully coordinated, districtwide math and science programs. During a child's school years, ideas should not be repeated while others are ignored. This issue alone points to the fact that reform is easier when taken on by whole schools and districts. Often, however, reform comes

---

1. My district is presently reviewing and revising these objectives to make them align more closely with state and national standards and "best-practice" research.

one teacher at a time. In either case, selecting objectives can be done using the following criteria:

1. Which objectives seem developmentally attainable? Knowing your students, especially their development at this time of year, and reflecting on your past experiences, what ideas seem within reach of most of your students?

2. Which objectives match available resources? For example, our school has an outdoor garden, so many life science objectives are addressed while studying plants. A school without this resource may rely more on animals in the classroom to gain life science knowledge.

Let's say that one of your units is on life cycles and environment. In order to select objectives, you might carefully consider the national, state, and local standards. These are listed in Figure 4–1, using my district's objectives (Glenview Public School District 34 1996), the State of Illinois benchmarks (Illinois State Board of Education 1997), and selected standards from the *National Science Education Standards* (National Research Council 1996). As I look at this list to use with my students, several issues come to mind. First, all the district objectives—with the exception of "describe cyclical behaviors of animals"—seem both developmentally attainable and possible considering my resources. Second, the Illinois K–2 benchmarks are on the topic of environment. The concept of life cycles is not recommended for study until grades 3–5. Although this could be a red flag for an age-inappropriate concept, in this case it does not seem to be a concern. Units taught earlier in my district seem to pave the way toward these topics, and none of the research cited in the national science standards suggests that life cycle concepts are beyond the grasp of second graders. Third, many of the national standard objectives seem appropriate considering both my student's developmental levels and my resources.

Therefore, I can select the objectives from the national standards shown in Figure 4–1, all of which fit my unit. I know that these are an integral part of a nationally constructed, research-based curriculum. They also overlap nicely with what I am required to teach by my state and district.

How Can Students Learn These Important Ideas?

With objectives thoughtfully selected, the next step is to plan how children will learn them. In other words, how will your math and science programs be structured, what activities are most likely to lead children to-

DISTRICT SCIENCE CURRICULUM GRADE 2	STATE OF ILLINOIS STANDARDS AND BENCHMARKS K–2	NATIONAL SCIENCE EDUCATION STANDARDS CONTENT STANDARDS: K–4
The learner will be able to	As a result of their schooling students will be able to	As a result of activities in grades K–4, all students should
• Describe the life cycle of plants and animals. • Observe and explain the life cycle of a plant by interacting with seeds and plant structures. • Conduct a simple plant investigation. • Create and interact with models of seeds and plants. • Observe and record changes in the life cycle of an insect. • Describe cyclical behaviors of animals. • Illustrate the interaction of plants and animals. • Explain how changes in the environment affect living things.	• Identify and describe the component parts of living things (e.g., birds have feathers; people have bones, blond hair, skin) and their major functions. • Describe and compare characteristics of living things in relationship to their environments. • Describe how living things depend on one another for survival.	• Understand that plants and animals have life cycles that include being born, developing into adults, reproducing, and eventually dying. The details of this life cycle are different for different organisms. • Understand that plants and animals closely resemble their parents. • Understand that organism have basic needs. . . . Organisms can survive only in environments in which their needs can be met. • An organism's patterns of behavior are related to the nature of that organisms' environment . . . When the environment changes, some plants and animals survive and reproduce, and others die or move to new locations. • Develop abilities necessary to do scientific inquiry: ask a question about objects, organisms, and events in the environment, plan and conduct a simple investigation, employ simple equipment and tools to gather data and extend the senses, use data to construct a reasonable explanation, and communicate investigations and explanations. • Understand that scientists develop explanations using observation (evidence) and what they already know about the world (scientific knowledge). Good explanations are based on evidence from investigations. • Understand that simple instruments, such as magnifiers, thermometers, and rulers, provide more information than scientists obtain using only their senses. • Understand that scientists make the results of their investigations public; they describe the investigations in ways that enable others to repeat the investigations. • Understand that scientists review and ask questions about the results of other scientists' work.

FIGURE 4–1   A comparison of local, state, and national objective for life science.

ward understanding, how can these activities be organized into workshops, and what framework for presenting the workshops is most effective?

ROLE OF WORKSHOPS IN THE MATH AND SCIENCE PROGRAM

Workshops are the primary delivery system of my math and science programs. I teach five math workshops a week, each lasting about an hour. Science workshops take up 2 to 3 hours per week.

Your beliefs, resources, and constraints influence your science and math programs. Many primary teachers realize that getting children to read is of the utmost importance and consequently devote the greatest amount of time during the school day to language arts. It is fortunate that in workshop math and science, spoken and written language are used frequently, and this of course supports language arts development. This offers a bit of flexibility in scheduling each subject. In some workshops, for example, children can write for 10 minutes in their learning logs. On these days it may be possible to reduce language arts time by that 10 minutes.

There are some situations when math and science is better taught through more direct instruction. For example, skills such as counting aloud by 5s need to be practiced until memorized, and this is best done through direct instruction. It is difficult to determine exactly what would be included on this direct-instruction list, but I suspect the list is short. As I work more with the workshop, some topics that I thought should be taught directly through drill and memorization (such as the addition facts) turn out to be more suited to the workshop. It seems that there is little in math and science that children don't learn better with as much hands-on activity, choice, reflection, and problem solving and inquiry as possible.

Activity Selection

Activity is at the center of math and science workshops. Let's say that you want your students to learn multidigit addition and subtraction or to meet the objectives you have selected for your earth materials unit. What activities are most likely to help? The best activities have these characteristics:

• *Good activities can be done hands-on.* When students are directly observing, interacting with, and manipulating phenomena, their learning and interest are likely to be greater than when teaching is less direct.

49

• *Good activities let children choose.* Choice in what children work on, how they work, and what they learn from the activity allows children to construct their own knowledge. Good activities help balance children's need for choice with their ability to choose wisely.

• *Good activities encourage reflection.* Learning is a social process. Activities in which children are able to interact with peers and their teacher are usually considered more enjoyable. More than the fun factor, however, is the fact that as students communicate ideas they often learn more.

• *Good activities allow for problem solving and/or inquiry.* These processes are at the heart of math and science. Addressing personally relevant problems and curiosities motivates children to learn. Additionally, problem solving and inquiry let children see math and science as useful.

• *Good activities can lead students to a wide range of understandings.* Because children are both at many different developmental levels and hold different interests, they are likely to come to a range of conclusions from an experience. The best activities are designed so that several different important concepts can be learned from the same activity.

• *Good activities can be approached in different ways.* Again, because children are different, they are likely to approach activities in different ways. Good activities allow children to use a wide range of strategies, from basic to sophisticated.

• *Good activities are open-ended.* Aside from avoiding the I'm done! phenomenon, activities without a definite end allow children to explore a topic as much as they care to. Also, much of math and science found outside of school is open-ended, so experience with this type of activity is useful.

• *Good activities are connected to life outside the classroom.* Often children are more excited about an activity if they see that it has real-life applications. Connections to experiences they have away from school help children see the usefulness of math and science. Children and adults perceive real life differently. The destruction of the rainforest, for example, may be considered by adults to be a significant problem, but it is not significant for many children—it is out of their range of understanding. Real-life topics for children are usually very local, such as the destruction of the tree in the playground.

• *Good activities address important math and/or science topics.* Activities can have all the above characteristics but still be worthless if they do not help children learn the significant ideas in math and science. There is no time in the school day for cute activities that do not go anywhere important.

Activity Types

Activities that fulfill the preceding requirements can be classified into teacher-directed and student-directed activities. The type of activity defines workshops during the activity period.

Teacher-Directed Activities   There are four types of teacher-directed activities: explorations, games, problem solving, and inquiries.

1. *Explorations:* Explorations are hands-on activities in which children explore math or science phenomena. Narrow objectives are not attached to these explorations. Rather, the goal is to give children the "big picture," to find out what children are curious about, and to get children excited about the topic. Two excellent sources detailing math and science explorations are *Elementary Mathematics and Science Methods* (Foster 1999a) and *Sciencing* (Phillips 1992).

   Examples: Explorations begin with things that children can see and touch and these simple instructions: Learn as much as you can about what you have there. As we begin to study geometry, students are given different sets of manipulatives likely to get them thinking about geometry. These include pattern blocks, mirrors, wood cubes and other 3-D shapes, and geoboards. Students examine, build with, and informally talk about their objects. As the children work, I observe and question children, hoping to find what children are interested in. The question "What are you doing?" posed to a first-grade boy building a pyramid out of wooden cubes (Figure 4–2) led to this conversation:

Student:	I'm building a tower.
Teacher:	Tell me about your tower.
Student:	It has this special one at the top (pointing to the sole orange block among the brown ones) then it goes down in order.
Teacher:	What do you mean "goes down in order"?
Student:	Well, below the special one there are four blocks in a square and then another square made out of nine blocks. I wish I had more of these blocks, then I could make another bottom.
Teacher:	You do have some more there. (Here I pointed to a pile of five blocks, which I knew were less than were needed to build a four-by-four base.)
Student:	That's not enough. I tried to build a bottom but I don't have enough. I wish that I had more.

51

FIGURE 4–2   Exploring geometry. *Photo by Daniel Heuser.*

Teacher: How many more do you need?

Student: (Considering) I don't know.

From conversations like this I sensed that building and planning geometric patterns fascinated many children. I left this exploration with a list of topics and questions from which I could plan a unit—topics that children were interested in and that were developmentally appropriate. Students gained a sense of what geometry was all about (building, describing, transforming, visualizing) and excitement about learning more about shapes.

Explorations are also used in science. We began our unit on earth materials by going out to our garden, where each child dug a small shovelful of soil into a bag to take back into the classroom. Back in the classroom, children were instructed to "dump the soil out on your desks and see what you can find out about it." For 45 minutes children were actively involved with their soil. Some squeezed it into balls and then broke it apart. Others made piles: leaves, roots, rocks, dirt, and bugs. While they worked, they informally exchanged observations with children around them. "Look at this worm trying to hide." "I have so many rocks in my soil." "How come you have so many bugs but I don't have any?" "My soil is so black." During the reflection period I asked children to write one thing that they noticed and one question they had about their soil. From these reflections (as well as from my observations during the activity period) I made the following notes:

1. Children resisted idea that rocks, sand, and humus were part of soil. Need to confront this misconception.

2. Three questions stood out:

   • What will happen when I do stuff (smash, put in water) to soil? (Fits nicely with unit objectives.)

   • Why were worms in some soil but not others? (Overlap with study of habitat—deemphasize)

   • What are the differences between different soils? (Good for observing/describing/comparing)

3. Students shared experiences they had digging in soil and playing in the mud. Need to connect these experiences with our work in school.

These notes formed the basis for planning the unit. Without this exploration, I could have picked topics that did not interest students or did not address their present beliefs about soil. Students' exploration of soil revealed a wealth of information for both students and teacher.

2. *Games:* Most children love to play games. Games allow them to socialize and fulfill their need for play. Games can be excellent activities to help children develop understanding in math and science, especially those that require children to reason, communicate, and mentally plan.

   Some of the best games for developing math reasoning and computation have been collected in the book *Young Children Reinvent Arithmetic* (Kamii 2000). A math game for teaching addition facts is the activity in the first sample workshop detailed in Chapter 5.

Examples: Games in which children sort objects can be used in many science units. Students touch and examine real objects in order to play these games. As a result, students can gain physical knowledge of the objects and develop classification skills as they engage in reflective debate with fellow players.

One such game is Rock Around. We play it as part of our unit on earth materials. Throughout the unit, students collect and observe rocks, and over time we compile a list of different rock properties, such as hard and soft, colored and clear, rough and smooth, simple and composite, etc.

This game is played in groups of three to four children. A wide variety of rocks (each fairly clear examples of the rock properties), cards on which are printed the rock property words, and a circle made of rope or plastic or drawn on paper are needed for this game. The first player secretly chooses one of the rock property cards and places it face down near the circle. She then chooses a rock from the pile and places it either in the circle or outside of the circle, depending on whether or not the rock shows the selected property. For example, if the secret card is *rough,* the first child would put a rough rock inside the circle or a smooth one outside of the circle. The second player then guesses the attribute. If the second child guesses correctly, he can keep the card. If he guesses incorrectly, the first child places a new rock inside or outside the circle, and the third player takes a guess. It then goes to the next player until someone wins the cards. Then the winner gets to choose the secret card, and play continues. The game is over when all the cards have been won.

More difficult variations of this game can be introduced if you notice that the basic game is too easy for some students. I have found that these variations can be modeled for the whole class, and then if children are allowed to choose which variation to play, they tend to pick the one that most matches their abilities.

In one variation, students can use two or three overlapping circles to form a Venn diagram (Figure 4–3). The child choosing the properties chooses two (or three) instead of one, and the other players have to guess both (or all three). In another variation, rocks can be placed not only inside or outside the circle, but on the circle as well. Those placed on the circle are not clearly identified with the selected property. In the example just given, a rock that has both smooth and rough portions would be placed on the circle.

3. *Problems:* In math and science education, the terms *inquiry* and *problem solving* are given different definitions by different educators, and these definitions often blur the distinction between the two. Rather than debating these definitions here, it is useful to look at these two processes as being on a continuum of choice. Who chooses the problem (or the question): the teacher, the student, or the teacher and student working together? This continuum is shown in Figure 4–4.

Another way to understand problem solving and inquiry is to see what they are *not.* First, a *problem* is different than an *exercise.*

FIGURE 4–3 Rock Around using a Venn diagram. *Photo by Daniel Heuser.*

Teachers often give students problems, but then show them how to solve those problems. Student completing these exercises are not solving a problem in the sense described in *Principles and Standards*—"engaging in a task for which the solution method is not known in advance" (National Council of Teachers of Mathematics 2000, p. 52). Instead, they are just getting exercise in applying the teacher's solution method.

Similarly, *inquiry* science is not synonymous with *hands-on* or *activity-based* science. While good science inquiry is primarily hands-

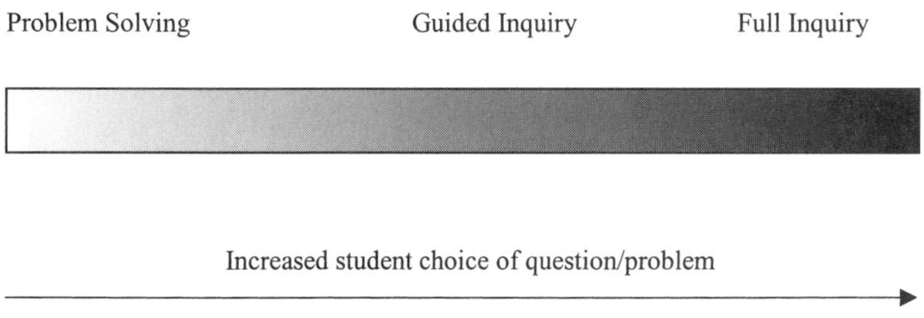

Problem Solving	Guided Inquiry	Full Inquiry

Increased student choice of question/problem

FIGURE 4–4 Problem solving and inquiring continuum.

on, inquiry requires more of students than just doing activity. According to the *National Education Standards* (National Research Center 1996), "(*f*)*ull inquiry* involves asking a simple question, completing an investigation, answering the question, and presenting the results to others" (pp. 1–2). In *guided inquiry*, teachers are more involved in the process, helping students to ask, investigate, answer, and present. For example, teachers may pick several questions for students to investigate that are based on students' interests.

Some districts have adopted hands-on science kits that include preplanned activities and materials. While these are a vast improvement over traditional hands-off instruction, it is only inquiry if activity is determined to some degree by children's questions. Two excellent resources define further the differences between hands-on inquiry and hands-on activity that is not inquiry-based. Both are available online: *Foundations: The Challenge and Promise of K–8 Science Reform* (National Science Foundation 2000), and *Inquiry and the National Science Education Standards: A Guide for Teaching and Learning* (National Research Council 2000).

When planning an inquiry or a problem-solving activity, you need to consider several issues. How able are your students to come up with their own inquiries that are practical, developmentally appropriate, and involve math or science that is part of your curriculum? It is unlikely that this will happen much before late first grade, but when students are ready, this is a powerful learning tool.

For much of the primary years, guided inquiry works better than full inquiry. Young children occasionally cannot think of topics (questions or problems) that are interesting, practical, and significant on their own. For this reason, it is important for teachers to guide students toward a topic, set the topic with input from the students, or come right out and declare "This is the topic!" Two examples of guided inquiries are included in this book: one involving tadpoles in Chapter 1 and another with plants that is detailed in the following chapters.

Not everything can be learned through inquiry. Students also have to solve problems that you give them. Not only does this help them learn important math and science, but children also need to know how to solve problems, just as they need to know how to inquire into their own curiosities.

1. *Problem solving with objects:* Problem solving with objects involves students solving teacher-generated problems involving objects. Using problems derived from objects offers different possibilities for student learning. First, because students cannot solve the problem without manipulating the objects, there is a tremen-

dous potential for students to develop cognitive structures. Second, we know that children love to touch, so interest is especially high in this type of activity.

The book *Little Kids—Powerful Problem Solvers* (Andrews and Trafton 2002) has detailed scenarios of children solving problems, most of them using concrete objects. A problem-solving activity involving estimating the quantity of objects is detailed in Heuser (2000a). Following is another example of this type of activity.

Example: How many grains of rice can fit in a baby-food jar? In this activity, jars are filled with dried beans of different sizes as well as rice. For example, five jars could be filled with lima beans, five with garbanzos, five with large kidney beans, five with split peas, and five with rice. Students are instructed to select a jar and find out how many beans are in that jar. Other materials needed are: construction paper to catch spills and graphically group beans, bottle caps, medicine-sized cups (available at drug stores), and a sheet on which children can record their work.

Children try different solution strategies. The simplest is to count by ones. This strategy works well for the larger beans because there are fewer of them. Very young children do not see the need to move each counted bean away from the uncounted ones and can lose count or count incorrectly. You can demonstrate good counting form to these children. More advanced children count by twos, fives or tens, moving the counted beans (in groups of twos, fives, or tens), away from the uncounted beans. A still more sophisticated strategy is to group the beans into piles of the same amount and then to count by that number. Some children will group them in quantities difficult to count (by threes or sevens, for example), whereas others see the connection between counting by twos, fives, tens, twenties, twenty-fives, one hundreds, etc., and grouping beans in those quantities. One child's recording strategy is shown in Figure 4–5.

An even more advanced strategy, especially for larger quantities, is to group the beans into smaller groups made of smaller subgroups. In other words, beans can be put into piles of 10, with 10 of these piles placed in a row or cluster away from the next group of 10 tens. Students then count by tens. For very large quantities, advanced students figure out that if you count the number of rice grains in a bottle cap and then find out how many bottle capfuls of rice are in the whole jar, that the total number can be approximated. Each strategy shows a progression of mathematical sophistication, and part of the teacher's job is to move children

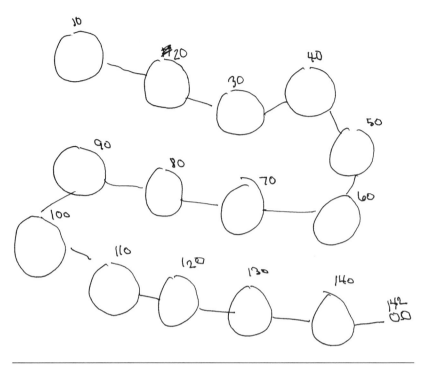

FIGURE 4–5   One child's method of keeping track of counted beans.

toward more complex strategies that they are capable of understanding.

In addition to counting and number sense, this activity can help students construct a number of other math ideas. Children can begin to see estimation as the most efficient alternative to counting numerous, small pieces. As students group and subgroup the beans,

1. They are developing and using their class inclusion abilities.

2. They can make connections to addition, subtraction, multiplication, and division. To use the 10 tens example, many links are possible: counting by tens is like adding tens, 9 groups of 10 is like 100 – 10, 10 groups of 10 is 100, 100 can be divided into 10 groups of 10.

3. They can make a connection to place value: there are 12 tens in 125, just as there are 125 ones, just as there are 1 hundred, 2 tens, and 5 ones or 1 hundred and 25 ones.

Although students can learn these ideas just by working alone with beans, teachers can make a tremendous difference by what they do during and after this activity. Teachers can question and probe during conferences. They can also have children work in pairs so that students can bounce ideas off of one another. During the reflection period, students discuss and write about what they have learned. All these behaviors help children construct these important relationships.

2. *Problem solving without objects:* Problem solving without objects is used mostly for math and superficially resembles traditional worksheets. One or more problems are written on paper and students work to solve them. Problems can be story-problem based (If you have 8 marbles and you get 17 more, how many do you have?) or they can be more abstract (8 + 17 = ?). Even though this type of activity is called problem solving *without objects*, "without objects" is a bit of a misnomer. Students can use objects to solve these problems if they like, as in the following example. However, the problems are not found in the objects, which is the criterion for the activity type problem solving with objects.

These problem-solving sheets are different from traditional math worksheets in several ways. First, they tend to have fewer, but more complex, problems. This is because students can learn more by thoughtfully solving five problems than by mindlessly solving forty. Second, the first problems are fairly easy, but problems become progressively more difficult throughout the sheet. The last problem may be solvable by only the most advanced students. This allows children of all levels to succeed and to be challenged. And because students rarely finish the whole sheet, you can avoid the I'm done! problem. Third, students are expected to write more than numbers for the answer. Lots of white space is provided below questions for children to make notes, drawings, diagrams, or graphs or for them to jot down a sentence. Lastly, questions often have more than one correct answer, which helps communicate a more accurate view of mathematics.

One excellent source for similar math problems is *Children's Mathematics: Cognitively Guided Instruction* (Carpenter et al. 1999). This book is rich in sample problems, but just as importantly it classifies story problems into types so that you can create your own problems. Most math textbooks also have questions that can be adapted to fulfill the preceding requirements.

Example: Figure 4–6 shows three examples of problem-solving sheets. Sheet (a) is designed for geometric reasoning and measurement fluency, (b) is to develop computational understanding, and (c) is to help students draw inferences from their tadpole experi-

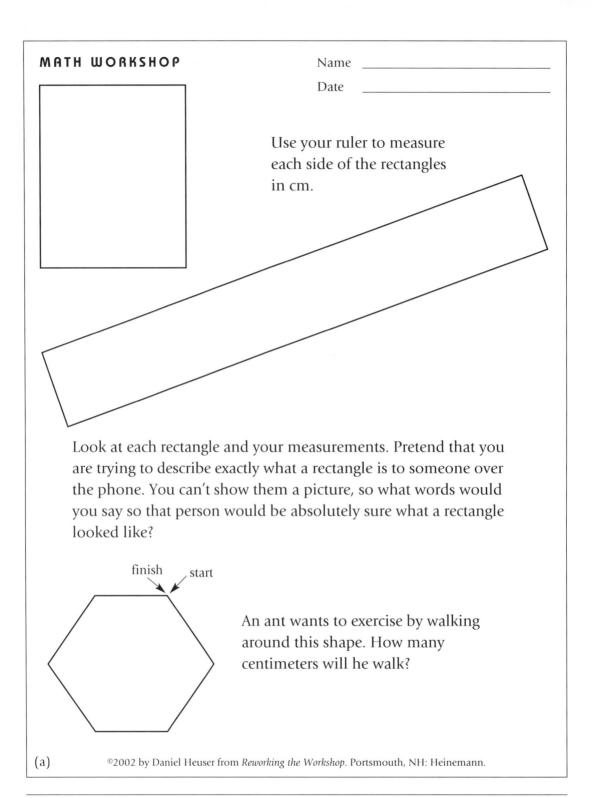

MATH WORKSHOP

Name _____

Date _____

Use your ruler to measure each side of the rectangles in cm.

Look at each rectangle and your measurements. Pretend that you are trying to describe exactly what a rectangle is to someone over the phone. You can't show them a picture, so what words would you say so that person would be absolutely sure what a rectangle looked like?

finish   start

An ant wants to exercise by walking around this shape. How many centimeters will he walk?

(a)

FIGURE 4–6   Problem-solving sheet.

MATH WORKSHOP

Name _____

Date _____

You are counting with your objects, and you put them into piles of 10. At the end you have 6 piles of 10 and one pile of 6. How many objects do you have all together?

Then, your partner takes away 2 of those piles of 10. How many do you have now?

You have 4 piles of 10 and 2 ones left over. How many all together?

Our turtle will cost $14. Mr. Heuser has $10. How much more money will we need to buy the turtle?

How many piles of 10 can you make out of 70 objects?

Solve.

10 + 10 + 10 + 10 + 10 + 10 + 10

(There are 7 tens.)

FIGURE 4–6   Continued.

Name _____

Date _____

What makes tadpoles grow faster?

		alive	dead
Water	shady	2	0
	sunny	2	0
	dirty	2	0
	clean	0	2
Kind of food	duckweed	4	2
	pellets	4	3
	lettuce	0	4
	fruit	0	2
How much food	a lot	1	0
	a little	1	0
Exercise	lots of room	1	0
	a little room	0	1

_____

How many alive tadpoles all together.

_____

How many dead tadpoles all together?

_____

How many tadpoles were in the "kind of food" experiments?

_____

What kind of food was the best? Why do you think so?

FIGURE 4–6   Continued.

Name _____

Date _____

Is it better to keep tadpoles in dirty water or clean water? Why do you think so?

Is it better to feed tadpoles a little bit of food or a lot of food? Why do you think so?

Pretend that you did the "dirty water or clean water?" experiment 10 times, because you wanted to be sure that dirty water was really better. If the results were the same each time you did the experiment, how many tadpoles would die?

Pretend that you are doing another experiment. Which tadpole home would be better? Why do you think so? Write on the back.

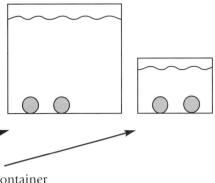

big container
two pellets

small container
two pellets

FIGURE 4–6   Continued.

Name _____

Date _____

Draw and label the perfect tadpole habitat.

Use all that you learned from our tadpole experiments.

What kind of water would you use?

What kind of food?

How much food?

What size container would you use?

FIGURE 4–6 Continued.

ment data as well as to practice using data from a table. Children are often assigned a partner of similar ability to complete these sheets, and the pairs are soon reading through the questions and discussing possible ways to solve them.

My goal during this type of activity is to spend as much time meaningfully conferencing with children. Answering questions such as What does number 3 say? and chasing after objects for children take away from this conferencing time. To avoid these problems, children are usually paired with a partner of similar ability. Not only does this encourage a productive exchange of ideas, but also it increases the odds that at least one of the partners will be able to read the question. Additionally, I read through the questions during the minilesson, so that even the youngest children know what the questions are. To avoid having to retrieve objects to help children model problems, I keep a number of plastic cups filled with counters. Students know that they can take a cup whenever needed.

As would be expected in an environment where choice is valued, many different problem-solving strategies emerge. Figure 4–7 illustrates how several children tackled one of these problems.

4. *Inquiries:* As discussed before, a child conducting an inquiry has some choice in the inquiry question, how to run the inquiry, and the conclusions that are drawn. Children can inquire into both math and science topics. Is coffee, water, or cola better for a plant? is the topic of a science inquiry detailed in Chapter 5. Two other science inquiries on the mixing of solids and liquids are described in Heuser (2000b) and Wafler (2001).

A Math Inquiry of Arrays: Arrays are rectangular arrangements of objects in rows and columns. They are rich in mathematical relationships. An egg carton, for example, is a 2-by-6 array that illustrates $2 \times 6 = 12$, $6 \times 2 = 12$, $12 \div 6 = 2$, $12 \div 2 = 6$, $6 + 6 = 12$, and $12 - 6 = 6$, among others. This inquiry was adapted from one designed by Diane Czerwinski, the gifted coordinator at my school.

This inquiry into arrays is best done over 2 days. The first day is spent in exploration. After briefly defining and demonstrating what arrays are, have students explore arrays by building the arrays with manipulatives (1-inch by 1-inch tiles work nicely) and recording them on graph paper. The teacher conferences with groups of children: How many tiles did you use? Can you make another array with that number? How many rows is that? Can you make an array with five rows? Why not? Can you make one that is longer and skinnier? Can you make a square?

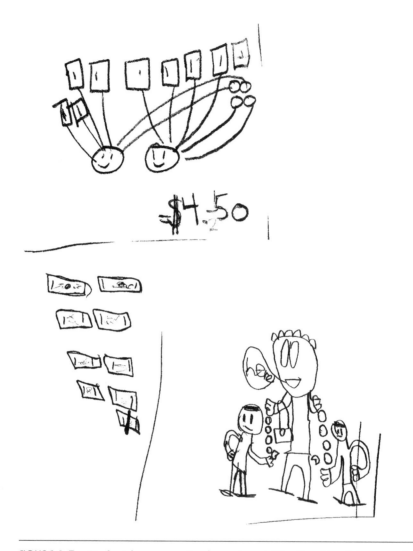

FIGURE 4–7 Students' responses to If you shared $9 with a friend, how much would each of you have?

As they explore 12 and other numbers, students begin to wonder aloud about things they find curious. Below are listed some questions that I overheard as a class of second graders worked with their arrays; I altered some of the questions to make them more likely to lead to substantial math. In the reflection period I had them choose one question that they would like to pursue in the next day's workshop:

1. How many arrays can you make out of 16 tiles?

2. What number of tiles will make a square array?

3. What number makes the most arrays?

4. With which number can you make the most arrays: 9, 10, or 11?

5. Can you make more arrays with even or odd numbers?

6. What arrays can you find around the room (windows, etc.)?

The second day begins with the teacher presenting something that she was curious about, and together with her students, conducting the inquiry. For example, I told the class that I noticed that only two arrays could be made with seven tiles: a one-by-seven and a seven-by-one. I was curious if there were any other numbers like that. How could I find out? Together we devised the strategy of starting with one, testing how many arrays could be made with that number, and then moving to the number that is one more. We would record our findings in a table. We carried this out together, using the numbers 1 through 13. Through this minilesson I modeled the inquiry process and motivated children to carry out their own inquiries, which they did in the activity period. During reflection children shared their questions, findings, and any new questions that they thought of.

Student-Directed Activities   Student-directed activities are wide open: children choose from a wide variety of object sets. They pick their materials, and work with the objects as they see fit. These activities take place in the context of the student-directed workshop. A sample student-directed workshop is discussed in Chapter 5.

ARRANGING WORKSHOPS

With a number of possible workshop activities in mind, it is time to arrange these workshops within the unit. Each unit generally consists of several series of workshops. For example, an earth science unit on earth materials might consist of two series: rocks and soil.

The Problem-Solving Framework

Most math workshop series (and occasionally science workshop series) consist of children solving problems. The problem-solving framework (Figure 4–8) is a way to structure each series. This framework organizes the workshops of each series into three different phases. Each phase generally includes two or more workshops.

A sample problem-based unit on the topics of the basic math facts is shown in Figure 4–9.

PHASE	WORKSHOPS FOCUS ON . . .
Invention	. . . problems that encourage children to develop their own understandings of the desired math or science ideas.
Development	. . . students communicating their personal understandings and debating the relative merits of each understanding. Occasionally, the teacher will present appropriate adult-accepted understandings of the topic, although children are not required to adopt these.
Reflection and Practice	. . . applying concepts and skills in a variety of situations, with the purpose of gaining confidence, becoming fluent, and developing more sophisticated understandings. Usually this includes extending knowledge into other real-world situations.

FIGURE 4–8   Problem-solving framework.

PHASE	WORKSHOPS FOCUS ON . . .
Invention	**Workshop 1:** Play "Roll and Grab"
	Minilesson: Model playing the game. Assign partners and pass out supplies.
	Activity Period: Partners play game. I monitor class for possible partner clashes, misunderstandings of the game rules, etc.
	Reflection: Students and I reflect on game logistics: ways of solving partner conflicts, misunderstandings about game rules, etc.
Development	**Workshop 2:** Play Roll and Grab, variations of Roll and Grab, and other addition games for as many workshops as students are interested and progress is significant.
	Minilesson: Tell students that playing this game will help them learn about numbers and addition. Go over some of the strategies brought up in previous reflection periods. Model playing the game, using those strategies. Assign partners and pass out supplies.
	Activity Period: Partners play game. I question children: Who has more? How do you know? How did you get that sum? What's another way of figuring it out?
	Reflection: Student share, in the large group, some strategies they might use to add dice pairs that I select. Encourage the use of more sophisticated strategies by asking questions such as, Is there a faster way to get this total?

FIGURE 4–9   Problem-solving sequence for addition facts.

Reflection and Practice

Workshop: Problem solving without object sheets.

Minilesson: Review the connection between the game and addition. Together, complete the first several problems, discussing strategies and making connections to the game. Preview the remaining questions, and assign partners.

Activity Period: Partners complete problem solving sheets.

Reflection: Student share, in the large group, their answers to several of the questions on the sheet. Strategies are shared and debated, and children are encouraged to try more sophisticated strategies that make sense to them. Then, individually, students solve a final problem (for example, If you baked 12 chocolate cookies and 5 oatmeal cookies, how many cookies did you bake all together?) as well as write their solution strategy.

FIGURE 4–9   Continued.

The Inquiry Framework

Science units (and some math units as well) follow the inquiry framework shown in Figure 4–10. This framework is a way to structure each workshop series. Like the problem-solving framework, the inquiry framework organizes the workshops of each series into three different phases. Each phase generally includes two or more workshops.

PHASE	WORKSHOPS FOCUS ON . . .
Exploration	. . . hands-on experiences likely to produce interest and knowledge of desired math or science ideas. From exploration, students produce questions.
Inquiry	. . . one (or several) inquiries based upon student questions. These inquiries include students sharing and discussing their results.
Reflection	. . . reflective activities to unify inquiries around the important unit ideas. Usually this includes extending knowledge into other real-world situations quite often through reading, writing, and discussion.

FIGURE 4–10   Inquiry framework.

To help illustrate the inquiry framework, one part of a science unit on life cycles and environment is detailed in Figure 4–11. Several of these workshops are expanded upon in Chapter 5.

PHASE	WORKSHOP PLANNING
Exploration	**Workshop:** Planting Seeds
	Minilesson: Show students dried pea pods. What is in each? Break them open to show seeds. What plants have you grown before?
	Activity Period: Pairs of students brainstorm what these seeds would need to grow into healthy plants. Each child plants two seeds.
	Reflection: Student share what their seeds will need, and a list is negotiated. Students draw and write in their plant journal about what they did, as well as write one question they have about how to take care of their seeds.
Inquiry	**Workshop:** Investigating Plant Needs
	Minilesson: Review list of plant needs and some student questions from earlier workshop. Ask if plants really need water, or will another liquid do? Explain how much people enjoy coffee and cola. Will these work for plants? How could we test this idea out?
	Activity Period: Class negotiates how to investigate this question. "Water" plants with coffee, water, and cola.
	Reflection: In journal, students draw experimental setup and write predictions.
	Workshops: Monitoring Experiments. This series of workshops may extend over several weeks, as long as interest is high.
	Minilesson: Observe plant progress and measure plants. In pairs, children discuss the results so far. What does it mean that one plant is bigger than the others?
	Activity Period: Students record plant progress through drawing and writing in their observational journals.
	Reflection: Students share their drawings, observations, and/or questions with the class.

FIGURE 4–11   Inquiry sequence for life cycles/environment.

Reflection   **Workshop:** Designing Other Plant Experiments

Minilesson: Present student questions from throughout the inquiry. Record other questions that children may have about plants and their environments. Have pairs of children select a question.

Activity Period: Pairs design an investigation to answer their question.

Reflection: Students present their investigation plans to the class. Whole group discusses strong elements and potential problems with investigations.

Workshop: People and Their Environment

Minilesson: Have class verbally summarize some of the ways that a plant's environment affects its health and growth. Ask how their environment affects people's health.

Activity Period: Students make posters showing positive and negative effects of the human environment.

Reflection: Students present and discuss posters. Then they respond in writing to this prompt: People and plants are affected by their environment. Pick one other living thing and write about how it is affected by the environment.

FIGURE 4–11   Continued.

CONCLUSION

As any experienced camper can tell you, having a plan doesn't necessarily mean that events will follow that plan. The series of workshops within a unit should be planned out ahead of time, but often something happens to alter what you originally envisioned would happen.

As an example, let me tell you about one year in my classroom, just as we were beginning a science unit on the states of matter. I noticed that students were coloring the tops of their plastic school boxes with markers during their free time and then pouring glue onto the box top. After the glue dried, they peeled off their creations: multicolored, flexible, dried-glue bookmarks. This became quite a craze. Students were fascinated by the different designs possible when they altered the type or amount of coloring and gluing. Lunchtime clubs were established to share this important information. I quickly realized that the main inquiry I had planned for this unit—What happens when different liquids and solids are combined?—was not the best inquiry topic for this crowd. So plans were changed, and that year students designed and conducted investigations to answer what really interested them: How can I make the best bookmark? Without a "big-picture" plan, teaching for understanding is nearly impossible. On the other hand, the original plan should usually bend to accommodate children's unexpected fascinations.

5

Three Workshop Sequences

The backgrounds and details of the main activities of three workshop sequences are provided in this chapter. The topic of the first sequence is the basic addition facts. The second sequence described is on the science topics of the life cycle and environments of plants. The third is on student-directed workshops. Later chapters describe how to teach effective minilessons to begin these workshops, what to do during the activity period, concluding each workshop with reflection, and assessing what students learned.

MATH GAME: BASIC ADDITION FACTS

Children need to know the number facts. Automatic recall of these basic number combinations allows children to fluently compute, and these facts come in handy both in school and at home. Aside from its mathematical value, knowledge of the math facts also is valuable on the public relations front. Children's mastery of the facts is often the preeminent criteria for the community to use in judging the success of its school's math program. In other words, if you can get the kids to learn the math facts, you are well on your way to having the parents on your side. The following workshop sequence shows one way to help your students develop automatic recall and understanding of the addition facts.

Background

The theories and activities of this sample math workshop are not mine alone. The work of Constance Kamii (1989, 2000) and of Thomas Carpenter and his colleagues (1989, 1999) has formed much of the theoretical and practical framework for the teaching and learning of addition.

Memorize or Think?   Children can learn math facts through two very different processes: memorization and thinking. Of course this is somewhat of a false dichotomy. Memorization requires thought, but considering memorization and thinking to be opposites helps to illustrate their differences.

Rote memorization has traditionally been the method used by teachers to teach addition facts. For example, 9 + 6 = 15 is learned by memorizing the sentence Nine plus six equals fifteen. These words are divorced from what 9, 6, and 15 actually mean. Typically, when teachers teach through memorization, students are drilled using flashcards and worksheets and were assessed through frequent timed tests.

In contrast to memorization, children can learn the facts by *thinking* about the numbers. For example, 9 + 6 can be solved by decomposing the 9 into 6 + 3 and then, using a child's doubles knowledge, solving (6 + 6) + 3. Or 1 can be taken from the 6 and added to the 9, turning 9 + 6 into the easier-to-solve 10 + 5. As children manipulate and reflect on number combinations over time, many (but not necessarily all) addition facts are committed to memory, to be retrieved without thinking. Many children, however, never commit every fact to memory but can automatically solve fact problems because they become so efficient with these strategies (Carpenter et al. 1999). To illustrate this, calculate 14 + 7. You probably knew the answer immediately, but I doubt that you ever sat down in school to memorize fourteen plus seven equals twenty-one. Instead, these thinking strategies have become so familiar to you that solving similar problems becomes second nature.

To build these computation strategies, it is necessary to embrace the informal problem-solving strategies that children learn in their years prior to school (Issacs and Carroll 1999). My four-year-old daughter, for example, can accurately solve a variety of computation problems, such as how many cookies she and her friend would have if she had 3 and her friend had 6. She often solves this type of problem by using her fingers, first holding up 3 and then putting up 6 more. This strategy is known as *direct modeling*. Finally, she counts all her fingers from 1 to 9. This strategy is called *counting all* (Carpenter et al. 1999).

Children learn other, more sophisticated strategies in several ways. First, when students are encouraged to describe and debate ways of solving problems, they often end up trying new methods. Second, problems can be designed that promote the use of newer, better thinking strategies. For example, when one addend is large and the other is small (such as 16 + 2), it invites the use of the more efficient *counting on* strategy—starting at 16 and counting on 2. Later, by discussing facts that are closely related to ones that children have already committed to memory, children begin to draw from what they know to solve unknown problems. Doubles such as 4 + 4, for example, are learned early on by children, and can be used to derive the answer to 4 + 5 (Carpenter et al. 1999).

Advantages of Thinking over Memorization   Should children learn the addition facts through rote memorization or thinking? There are several problems with rote memorization. First, children who memorize 9 + 6 = 15 don't necessarily understand what this means. We've all seen children have trouble with word problems that can be solved using simple addition facts that they can recite. Second, memorizing the facts robs children of one of the best ways to develop other mathematical concepts. As illustrated next, when children think about addition they develop a greater understanding of number than those children who just memorize. This is not surprising because these children spend their time thinking about numbers, whereas the other group is *not thinking* about numbers. Results of research illustrate the advantages of having children think about the addition facts.

Early Research   Three early studies have especially influenced thinking on how children learn facts with understanding. Brownell and Chazal (1935) found that frequent drill hindered children from advancing to more sophisticated strategies. For example, if a child counts 4 fingers on one hand and 5 fingers on the other to solve 4 plus 5 and then counts all of them one by one, frequent drill reinforced this strategy but did not help that child develop better strategies. In a later study, Brownell (1944) found that children who used strategies to solve facts had far better immediate recall at the end of the school year than students who simply memorized the facts. The third study showed that children who could explain how they solved basic facts at the end of the school year forgot far fewer facts over summer break than did students who could not describe how they solved facts (Rathmell 1978).

Kamii's Data   More recent data collected by Kamii (2000) further demonstrate some of the advantages of promoting thinking strategies. Her study compared children in a strategy-based first-grade classroom, with those in classrooms that taught the facts using a textbook and workbooks. In timed interviews at the end of the year, children who learned the facts by playing math strategy games, solving word problems, presenting and debating strategies, and using situations outside of math class (such as taking lunch count), outperformed those children encouraged to memorize in 26 of 29 addition facts. There was no difference between the groups on the other three facts. Usually the differences between the two groups were quite dramatic. For example, 52% of the strategies group could say that the answer to 5 + 8 was 13 within 3 seconds, compared to only 14% of those children who memorized.

 More than just a better ability to solve addition facts, the children in the thinking group in this study were able to apply their knowledge to solve word problems that required application of math facts. The textbook and worksheet group was much less able to use their facts to solve problems. This study supports much of the research found in Chapter 2: memorize-and-repeat teaching methods often sacrifice understanding for simple recall. And in this case, parrot math did not even yield better recall.

Math Workshop Project Data   Another study shows further advantages of thought-based instruction. Beginning in the 1997–1998 school year, I directed the Mathematics Workshop Project, a 4-year study of first- and second-grade students in two schools in my suburban Chicago public school district. One hundred eighty students were given a battery of interviews testing logical development and mathematical understanding during the study. By comparing students taught through a math workshop approach with those taught in other ways, I hoped to learn more about how the math workshop worked as a teaching technique. The study yielded a wealth of information, some of which has been summarized previously (Foster 1999b; Heuser 1999, 2000b; Heuser and Foster, in preparation). I will provide additional portions of the project data to illustrate certain points, starting with this chapter.

One of the interviews assessed children's knowledge of number families. A number family consists of all of the whole number pairs that, combined, equal a particular number. The 9 number family, for example, includes the pairs 9 and 0, 8 and 1, 7 and 2, 6 and 3, 5 and 4, 4 and 5, 3 and 6, 2 and 7, 1 and 8, and finally 0 and 9. Children gain this knowledge of 9 over time, as they physically construct and divide sets of objects and mentally manipulate numbers. Because it makes sense that a child with such intimate knowledge of 9 could fairly easily solve problems such as $6 + 3 = ?$, $6 + ? = 9$ (as well as $9 - 6 = ?$ and $9 - ? = 3$), we assessed children's automatic recall of the 4, 9, and 13 number families.

The number families interview was adapted from the *hand assessment* developed by Dale Rubley Phillips (1991). In each interview, the interviewer held out an open hand in which there was a given number of chips and a closed hand hiding an unknown amount. Having been told the total number of chips, the student had to figure out how many chips were in the closed hand. The interviewer then redistributed the chips and asked the same question. If a child correctly named, within 5 seconds, the hidden quantities in three of four trials, he or she was awarded one point and moved up to the next higher number family. Thus, a child who failed the 4 number family assessment was awarded 0 points, and so on, up to a maximum of 3 points by knowing the 4, 9, and 13 families.

Second-grade students in classes using the math workshop and those in a text-based program were given the assessment as a pretest in September. Those awarded three points were disqualified from this portion of the study. In May the posttest was given. Posttest results are shown in Figure 5–1.

GROUP	N	0	1	2	3	AVERAGE
Workshop	32	0 (0%)	1 (3%)	17 (53%)	14 (44%)	2.41[*]
Non-workshop	23	2 (6%)	4 (12%)	10 (43%)	7 (30%)	1.96

[*]Results were analyzed using analysis of variance (ANOVA). The workshop group mean was significantly higher at the $p < .05$ level.

FIGURE 5–1   Posttest results for number families.

The workshop group, despite starting the year with essentially the same average as the nonworkshop group, ended with a significantly greater knowledge of these number families. What is especially telling is that the control group was taught by teachers who used a program—Everyday Math (Everyday Learning Corporation 1989)—that many consider quite progressive.

Why the math workshop group did better is open to speculation. One likely possibility is that teachers following the text to the letter would be missing many opportunities for using the four elements that research shows helps children learn—hands-on activities, choice, reflection, and inquiry and problem solving. I have found this to be a common problem with many texts; the bones of worthwhile activities are there; they are just not fleshed out enough to reach their full potential. The next section will describe an addition facts game that is the main activity for the addition facts workshop sequence.

Addition Facts Game

The game is called Double Roll and Grab. I have found it worthwhile for first- and second-grade students of all abilities; kindergarteners that can count will also benefit. Students have the opportunity to develop their counting abilities (by ones, twos, fives, tens, twenty-fives, one-hundreds, etc.), and place value as well as the addition facts.

Game Rules   Four standard dice and between 25 and 100 counters are needed for each pair of students. Each player rolls two dice at the same time. The player with the higher total takes that number of counters from the bank and puts it in his or her winner pile. The other player takes no counters. If both players roll the same total, they both take that number of chips. Play continues until all of the counters in the bank are gone. The player with the most number in his or her winner pile wins the game.

Variations   Variations of this game can be played in order to meet the needs of your students and to encourage new, more efficient strategies. Most variations rely on different dice. Standard dice are used in Double Roll and Grab. These are the easiest because of the low numbers and because students can count the dots if needed.

Single Roll and Grab   Only one standard die is used. This variation is good for many kindergartners or students just beginning to count.

Special Double Roll and Grab   Students are given the choice (or guided by the teacher) of different nonstandard dice. These dice offer many opportunities for children to develop more advanced thinking strategies. Numeral dice, including those with numbers greater than six, can be used for more advanced students. Combining one numeral die with one standard die encourages children to use the counting up strategy, because they can start with the numeral (5) and then count up using the dots (6, 7, 8). This is especially true if the numeral die has higher numbers. Using two numeral dice requires children to add the numbers mentally and should be reserved for those students who are ready for them. If they are allowed to pick these dice before they are ready, most collect the chips separately (i.e., collect 5 and then 6), which has limited benefit.

Triple, Quadruple, etc., Roll and Grab   Increasing the number of dice encourages children to look for number combinations that are easier to add. For example, if a 6, 3, and a 4 are rolled, students often combine the 6 and 4 to make 10, and then add the 3. There are also times when children can decompose larger numbers. For example, if an 8, 9, and 14 are rolled, a child could solve it this way: $14 - 1 \Rightarrow 1 + 9 = 10$, $(14 - 1) - 2 \Rightarrow 2 + 8 = 10$, so $10 + 10 + 11 = 31$.

SCIENCE INQUIRY: PLANT LIFE CYCLES AND ENVIRONMENT

Background

Cycles can be found throughout math and science. Constancy and change, geometric and number patterns, cycles of rocks and water, time and seasons, the flow of energy, behaviors of living organisms, including people; knowing each of these is dependent on understanding cycles. The life cycle—birth, maturity, reproduction, and death—is at the same time abstract and immediate. Young children often are intimately familiar with events such as the birth of a sibling, the death of a pet, and the age-related differences between their grandparents, parents, and themselves.

At the same time, children have a hard time gaining the big picture of life cycles, in part because it can take many years for humans and animals to cycle through their lives.

Teaching Using Animals and Plants

One way to gain this understanding is for children to observe animals and plants that have a relatively short life cycle. In schools frogs and butterflies are often observed, partly because, as organisms that metamorphose, their life cycles are particularly striking. Plants such as green beans and peas are easy to care for and relatively quickly sprout, flower, develop seeds, and die.

Even with these specimens, understanding a life cycle is difficult for young children because the cycle is still progressing at a timetable that is usually longer than children's attention. Teachers need to be very vigilant in helping students connecting one life cycle phase to the next, summarizing what happened in previous lessons, and pointing out relevant details. A discovery approach toward teaching will not work here.

This was illustrated to me several years ago as I planned a second-grade unit of life cycles. I found from a pretest interview (an updated version is included in Chapter 9) that many of my students did not know that fruits are developed in flowers and that seeds can usually be found in fruit. I felt that these details were important pieces of the life cycle, so I planned an activity that at the time seemed foolproof: to learn about life cycles, students must see a life cycle.

And that's what we did. I brought in a green-bean pod from my garden that had dried. We took out the seeds and planted them in a half-barrel in the playground, and soon sprouts emerged. Through early September, the sprouts grew into vigorous plants, which in turn flowered. Before the first freeze hit, we saw the flowers fade, and in their place, tiny beans grew. Students recorded the cycle in sketches in their observation journal. Each drew the seeds in the dried bean, the tiny sprout, the full-grown plant in flower, and the fading flower giving way to the bean. The cycle was complete—in my eyes, at least.

For the posttest assessment, I had each child draw the life cycle of the green bean, from seed to seed. To my surprise, only a fourth of the students who originally didn't know on which part of the plant fruit and seeds originated knew now. Most of the students, the same ones who had accurately sketched the seeds in the pod just a month before, showed the seeds underground among the roots or encased in the branches. Almost no one drew the bean growing from the flower.

One likely reason for this is that I did not push the children to reflect on what they were seeing. As the children sketched I was running a reading group; I never had the opportunity to question them or to encourage them to verbalize their observations to their peers. The class as a whole never shared their sketches or was asked to use them in any way. We never discussed the plant's progress: how it had changes since last time, if they had ever seen other plants flower like that, or even, Where did we get those seeds from again? By not encouraging students to reflect on their observations so that I could emphasize important elements, many details never made it beyond the observation journal and into the students' understanding.

Raising animals or growing plants offers great opportunities for children to learn about the environments of organisms and to engage in inquiries, in addition to learning about life cycles. I have found that many students have questions that touch upon the environment of living things, such as what they are fed and what their habitat is. Often the best way to answer these questions is through hands-on inquiry. Studying an organism's environment at the same time the life cycle is being observed allows students to stay involved in important hands-on science while the life cycle slowly unfolds.

Plant Inquiry

One of these hands-on inquiries, involving pea plants, is described here. I related a similar set of inquiries on the frog life cycle in Chapter 1.

This activity has three main goals, summarized from the objectives listed in Chapter 4:

1. developing students' knowledge of the inquiry process

2. teaching that a plant's environment affects its health

3. increasing children's knowledge of the plant life cycle through direct observations

An Inquiry into Plants in Different Environments   Will plants grow better if they are given cola, water, or coffee? This was the inquiry question being explored by a group of two multiage (first- and second-grade) classes. This inquiry took place in the middle of our life cycles and environment unit. Before the actual inquiry, during the exploration phase of the inquiry framework, students each planted two pea seeds in paper cups. I had them plant two so that at least one would sprout for children to take home and so there would be enough second plants for us to use in an experiment. Students observed and discussed their plants' progress as they wrote questions they had.

Many of the questions revolved around plants' needs: Is my plant getting enough sun? Will my plant grow without water? What if I gave my plant milk? Even though no children asked specifically about the relative benefits of cola, water, and coffee, this question was both practical and likely to lead to significant science and played off a vein of the children's curiosity. Because the inquiry question was formed both by students and teacher, this inquiry would be a guided inquiry.

We began the inquiry phase of the inquiry framework when many of the plants grew to a height of about 6 inches. In the minilesson of this workshop, I reminded students our observations and discussions about their plants. I asked if plants really need water or if some other liquid would do. I suggested that coffee or diet cola might work just as well, because coffee makes some adults wake up and many kids love pop. Through a process of negotiation we devised a way to fairly test this question.

Black coffee and diet cola were the liquids of choice aside from water for two reasons. First, children are somewhat familiar with coffee and cola and they know that plants drink water. Second, I had let children pick their own liquid to test during the previous year. One of the unforeseen conclusions from this inquiry was that anything with sugar in it—juices, milk, nondiet pop—makes the soil incredibly smelly.

The activity period was short. I simply poured the three liquids into the soil of the specific plant and had children sketch the experiment set up (Figure 5–2). In the reflection period several children shared their predictions with the whole group. Finally, each child wrote down his or her prediction with an explanation. Over the next two weeks, we had five workshops in which we monitored and discussed the plants' progress. Finally, as it became obvious which plant was doing better, we negotiated our results.

Solving Problems on the Plant Life Cycle   As the inquiry progressed, I presented several problems to the students that were designed to help them better understand the plant life cycle. These activities focused on the areas that the pretest interview showed needed attention: that seeds develop in fruit and that fruit forms from flowers.

In one of these workshops, I asked students where seeds could be found on plants. Many thought that fruits could hold seeds, but quite a few students felt that seeds would also be located in the roots, stems, flowers, and leaves of the plant. These misconceptions are common in children. Technically, all flowering plants form seeds in a plant's fruit, which grows at the base of the pollinated flower. It is unfortunate that we also use the term *fruit* to describe sweet, edible plant parts such as apples and bananas. Because we do not think of green peppers or cucumbers as fruit, much less walnuts, wheat, or dried flower heads such as cotton, understanding that all seeds form in fruit is beyond the cognitive abilities of most young children. To address this issue, throughout the unit we used the term *seed holder*, a term that encompassed "fruits," seeded vegetables,

FIGURE 5–2   Experiment setup.

nuts, dried flower heads—anything that holds seeds. With this adaptation, children were ready to solve problems designed to facilitate understanding of the plant life cycle. Several of these activities will be presented in the next two chapters.

STUDENT-DIRECTED WORKSHOP

Background

Student-directed workshops are based upon the work of Darrell and Dale Rubley Phillips (D. G. Phillips 1992; D. R. Phillips 1991; Phillips and Phillips 1994, 1996). As part of their effort to translate Piaget's research into classroom applications, they proposed that students need to have greater control of what they are learning and how they learn it. Phillips and Phillips felt that if children were allowed self-directed sessions where they could follow their own interests while working hands-on with their choice of objects, they would be drawn to activities that were appropriate to their cognitive development. Children who interacted with different math and science manipulatives, such as pattern blocks, seashells, colored water, and mirrors, would have excellent opportunities to learn mathematical and scientific concepts and processes for which they were developmentally ready. Another benefit of this object exploration is that children would be likely to develop the cognitive structures—including classification, ordering, and conservation—that are the foundation for understanding math and science.

Object Exploration and Classification

To see how this would look in my classroom and to test the effectiveness of object exploration on children's classification development, I conducted another study involving three first-grade classes (Heuser 1996, 1997). Two classrooms formed the object-exploration group. The teachers of these classrooms conducted 20-minute periods, twice weekly, in which children worked with objects. This activity was usually in lieu of paper-and-pencil math exercises from the district-selected textbook. No formal instruction in classification was given.

The second group was the control. They did not schedule free exploration of objects but were taught math through the district math program, which consisted of workbook exercises and games. One lesson during this time was devoted to classifying attribute blocks according to different attributes.

Each child was given a one-on-one task interview (adopted from Phillips and Phillips (1996)) at the beginning of the study. Children were put in one of three categories of classification ability based on these pretest results:

Level A: Student could not group objects using a valid classification scheme (i.e. group by one defining attribute, such as size, shape, or color).

Level B: Student could group objects using only one valid classification scheme.

Level C: Student could group objects using two valid classification schemes.

After 4 months, students who tested at Level A in the pretest were given the same interview for the posttest. The results are shown in Figure 5–3.

GROUP		A	B	C
Object exploration	36	7 (19%)	15 (42%)	14 (40%)
No object exploration (control)	18	7 (39%)	11 (61%)	0 (0%)

FIGURE 5–3   Posttest results for classification (sorting).

Although most of the children in both groups could now classify by one classification scheme, the only children who could then go one step further and reclassify the objects based on a new set of criteria were those in the object-exploration group.

I drew two conclusions from this study as one of the object-exploration teachers. First, the data suggest that children given time to work hands-on, in ways that match their interests, are more likely to develop more sophisticated classification structures and thus be better prepared to understand math and science content. These results are not surprising given how children spend their free time. Video games, TV, and organized sports, although valuable in their own right, do not allow kids to work hands-on with collections of objects. It appears that giving children time in school to work hands-on gives them an advantage.

Second, the children I observed during these exploration periods were generally very focused, calm, and content. They had a sense of ownership in what they were doing. Children carefully examined their objects, looking at them from different angles, rubbed them against their cheeks, even smelled them. In addition to classifying, with no prompting from me, students naturally engaged in a range of behaviors. They made designs, stacked, counted, put objects into order, compared, weighed and measured, and talked about and made drawings of their projects. These sessions looked remarkably like . . . play!

Take these free exploration sessions and place them in the workshop format and you have a student-directed workshop (I use the term *free workshops* with the children). These are a part of my first-grade math and science program; student-directed workshops take place about once out of every five math and science classes. Kindergarten teachers tend to use these workshops more frequently. It has been my experience that many children start to lose interest in free exploration toward the middle of first grade. Children at that point of the year are often beginning to crave the more formalized instruction of teacher-directed workshops. On the other hand, some students enjoy and seem to benefit from student-directed workshops well into second grade. I offer this free exploration as one choice (the other choices being math games) during some workshops to meet the needs of these students.

Colored chips	Pattern blocks
Base ten blocks	Rulers
Plastic animals	Graph paper
Coins	Plastic cups
Inch cubes	Dice
Geoboards	Tape measures
Unifix cubes	Calculators
Seashells	Shape templates
Pebbles	

FIGURE 5–4   Useful object and tools for student-directed workshops.

Conducting Workshops

Student-directed workshops revolve around sets of objects. In my class-room, these sets are stored in clear plastic tubs for easy identification and access by the students. A list of some of the most useful objects is shown in Figure 5–4. There should be enough sets for every child to have one. For the most part, I have found that children engage in productive math and science behaviors more while working by themselves. More often than not, partners become involved in disruptive behaviors (such as play-ing "dinosaur war") rather than counting, classifying, ordering, experi-menting, or other important math and science processes.

These workshops begin with a minilesson. Student-directed mini-lessons are discussed in depth in the next chapter. Beginning with the first student-directed workshops, the workshop rules (see Figure 5–5) are em-phasized and periodically reinforced. It is important to set high expecta-tion for children in regard to taking care of materials, being respectful of others, and working hard. I always emphasize that, even though it may

1. Find your own space to work.
2. Work quietly so that you do not disturb others.
3. Be careful not to break or lose any pieces.
4. When you are done, put your collection away where you found it.
5. Work hard, challenge yourself, and enjoy your learning.

FIGURE 5–5   Workshop rules.

feel like they are playing, they have the responsibility to learn as much as possible. We never refer to these workshops as play. When I confer with the children (Chapter 7), I often ask, "What are you *working* on?" or "What are you learning?"

In the activity period, students make their selection of object sets, find a workspace on the floor or desk, and begin to work. I keep a list of students so that each student gets to pick first at some time during the year. While they interact with their objects, I walk around and conference with students. During these conferences I question children, make suggestions, pose problems, or offer encouragement. Occasionally I do informal assessments—these are discussed in Chapter 9. Student-directed workshops end with a reflection period (Chapter 8) and students putting away their objects.

6

The Minilesson

Each workshop begins with a minilesson. The word *mini* means that minilessons are shorter than traditional lessons, but duration is only one difference. Traditionally, a lesson was when the teacher tried to transfer knowledge to his or her students by telling them what they needed to learn: You use a ruler by holding it at one end of the object and then read the number at the other end. Let me show you. We now know that memorize-and-repeat teaching does not work very well, so the focus when the teacher talks to the whole class has shifted away from transferring knowledge and toward preparing students to build their own knowledge. Because knowledge is best built with learning that emphasizes hands-on activity, choice, reflection, and problem solving and inquiry, the lesson is shrunk in length to allow time for these elements in the later parts of the workshop.

WHAT DO MINILESSONS LOOK LIKE?

Most minilessons last about 10 minutes. The idea is to provide the necessary information and then get students off to work as soon as possible. In this short amount of time, the teacher usually does three things: identifies the main idea of the workshop, activates interest and prior knowledge, and explains the activity that will be done in the activity period. These three points are explained in the following example.

TRIPLE ROLL AND GRAB

The minilesson to introduce the addition game Triple Roll and Grab begins like most other minilessons. In my classroom, the children are seated on the floor in front of the white board. This is my preferred setup, but any arrangement will work that gives you a place to talk and write and that has limited distractions. I then plunge into the minilesson:

- *Identify the main idea.* "Today we are going to continue working on the addition facts."

- *Activate interest and prior knowledge.* Sharing situations in which children and adults might use what they are learning outside of school can motivate children. Say, for example, "I just used an addition fact this morning when I packed lunches for me and my wife. I like to have three cookies every day, and my wife said that she wanted two, so I went to the cookie jar and took out five cookies: 3 + 2 = 5." Pair shares, in which students briefly share with a partner, can be used to help students think of related experiences: "Everyone turn to a neighbor and tell them one time when you used an addition fact." The use of pair shares is detailed in Chapter 8. You may also want to make connections to past lessons: "Yesterday you played Double Roll and Grab. Today we will try a game that is similar, but you get to roll three dice!"

- *Explain the activity.* What, specifically, will students be doing next? What do children need to know to perform the activity? "The game is called Triple Roll and Grab." Modeling the activity is usually helpful. "Can I have a volunteer to come up here and play the game with me?" The minilesson is also the time to communicate rules and procedures. "In a second I'm going to announce partners. As soon as I call your name, get the dice and chips from the table over there, find a good spot, and start to play."

Keep in mind that these steps are not meant to be a script; the type of activity, level of students' attention, time availability, and other factors will affect what happens in minilessons. For example, late on Friday afternoons my students sometimes have little tolerance for hearing me talk. At those times my minilesson can be as brief as, "Let's play Roll and Add. Here are the materials. Any questions?"

PLANT INQUIRY

Just because minilessons are short doesn't mean that they can't be interesting. Children tend to remember novel presentations, so my teaching partner and I kicked off the plant inquiry together, drinks in hand. I grasped a Starbucks cup, enjoying my morning coffee, and Allison held a can of diet cola. Together, we identified the main idea of the workshop, activated interest and prior knowledge, and explained what students would be doing in the activity period.

Allison began: "You have been growing your own pea plants. Some of you asked if water was the best drink for plants. What do you think Mr. Heuser?"

"Well, water is tasty, but sometimes I really like a cup of coffee." I replied. "It really gets me going! I bet that coffee might make a plant grow faster."

"Coffee is good, but pop is delicious! Maybe this diet cola would be best for a plant," said Allison.

By now we had the children's interest. The room was already buzzing with conversation, so to keep things orderly we initiated a pair share: "Everyone turn to a neighbor and tell them which you think would be best for your plants, coffee, water, or cola." We also wanted to draw on any knowledge that children already had about plants and liquids, so we added: "Be sure and tell your partner *why* you think what you think! What do you know about plants, water, coffee, and cola that helps you make your prediction?"

We then identified the main idea of this workshop: "Let's do an experiment to find out which liquid is better! How could we test this out?" Together we negotiated the experimental design. Students suggested that we "water" some of the plants with coffee and some with diet cola to see what happens. We asked, "How will we know if they are better than water?" and they responded that we give some other plants water.

Together we defined the question: What will make pea plants grow taller, coffee, water, or cola? One child compared this to a race, and the children latched onto that idea. We used this scenario to introduce the idea of controlling variables. "Since we are having a race, would it be fair if we use these plants?" I asked, showing them three plants of different heights. After some discussion, the children agreed that that would not be a fair test. Allison and I also suggested that we put two of the plants right next to a window and the third in a dark spot in the corner. Again, students objected that this would be unfair.

Following the students' direction, we set the plants up as in Figure 5–2. The workshop concluded with an explanation of the activity period activity. "In your plant observation log, draw and label this setup." Setting expectations is also part of a minilesson. In order to make it explicit what a quality drawing would include, we added: "Last time some people forgot to label the important parts in their drawing. Scientists have to be very clear with their drawings, so make sure to include labels for the pots, plants, sticks, and signs."

Role-playing such as this is just one example of teachers using their strengths to give minilessons a special twist. If you are good at telling stories, singing songs, or drawing pictures, these techniques can be used to deliver minilessons. Remember, however, to keep minilessons mini. Too much "song and dance" may limit the time available for activity and reflection.

STUDENT-DIRECTED WORKSHOP MINILESSONS

The First Student-Directed Workshop

Ideally, the first "free" workshop of the year should be orderly, with clear, high expectations set for student behavior and with students knowing that they will be involved in important learning—even though it will probably be a lot of fun. These ideas are communicated in the minilesson.

This first minilesson can begin with a discussion of collections. Children often have collections at home, and asking a few students to share what they collect can lead into sharing your collections: "I want to show you some of *my* favorite collections. These are some blocks that connect, like this. I like to put them together and see how long I can make it. I also have plastic farm animals. See how some are bigger and some are smaller?" You can capture interest and lead students to think about how the sets could be used by showcasing a few of the sets like this.

Children need to know the purpose of student-directed workshops. It's important to communicate that what they are about to do may be fun but that it is a serious learning enterprise. It is also important that students understand that it is their responsibility to get the most they can out of student-directed workshops: "A lot of people don't know this, but when kids work with collections they can learn so much. What are some of the things that you can learn by working with, say, this collection of seashells? Good, you would definitely learn about numbers if you counted them! Some people might think that you are just playing when you are working with these collections, but we know that if you choose objects that you find interesting and work hard, that you can learn a lot. That is your job."

Next, present the workshop rules and the rationale behind the rules: "These collections are very important to me, and I don't want any to get ruined or lost. I want you to work with them too, but I have a few rules that you have to follow. I wrote them down here (Figure 5–5). Let me explain them to you now."

Clean-up time is especially difficult for some students, so it doesn't hurt to remind students several times about this procedure: "Remember, when you hear the bell, it is so important to immediately stop your work and begin to clean up. This is the hardest part for some kids, but I know that you can do it!"

At this point, students should know what to do with their objects. They now need to know the procedures for getting objects. "In a minute I am going to start reading off the names of the partners. I have all your names on this list so that all of you will have a chance to go first at some point. That way it is fair. As soon as I call your names, go pick a collection, find a space together, and begin your work." Students should now be ready to begin their first free-workshop period.

Minilessons on Management Issues

Minilessons for student-directed workshops often focus on management for the first weeks of school. Students need to learn how to take care of the objects, how to pick a good workspace, and how to record their work and reflect. Topics for these minilessons sometimes come from past problems. "Remember last time when we couldn't find the seashells because someone put them behind another container? Let's go over again how these containers are arranged."

Later Minilessons

When students are familiar with the rules and procedures of the student-directed workshops, the focus of the minilesson shifts to introducing new objects and activities.

Introducing New Objects   Children like things that are fresh and new, so occasionally minilessons can present object sets that are newly purchased or borrowed or that the teacher feels are underused. For example, you can show containers of clay that the children have not seen and ask, "Look at this clay. What could you do with this?"

The students have many suggestions: "You could build a person. Make a long snake or a pancake. See how flat you could roll it with a block." It is good to model each of these suggestions, making it clear that these are just a few things that students could do: "Wow, I made a pretty long snake. How long do you think it is? As long as your arm? Maybe you could try to make one even longer. Or try something else—you know best what is interesting to you."

Suggestions that are not so worthwhile, such as, "You could poke it with pencils," also need to be addressed. "That would be interesting, but I'm afraid the clay would get dirty, and that would be breaking one of our rules. Most kids don't want to pick the clay and discover that it has pencil marks all over it."

Introducing Activities   It is helpful to devote frequent minilessons to showing worthwhile activities that you noticed children doing in previous workshops. For instance, perhaps you noted during the last student-directed workshop that a boy was putting a set of bolts in order by length. If you feel that this would be developmentally appropriate for some of the children to try, present this idea in a minilesson.

"Last time, I noticed that Grant was putting his bolts in order, like this." Here you can show the bolts in a pocket chart, ordered from shortest to longest. "Can someone explain the pattern that Grant used here?" To extend this concept of ordering into other objects, ask, "What other objects can be put in order like this?"

If someone suggests that Cuisenaire rods could also be ordered, you can pull them off the shelves and place them in the chart, slightly out of order. This can create dissonance among those children who may be ready to order but are not yet secure in this structure. "You're right! I put them in order."

"NO!" reply some of the children, for whom this error was obvious.

"You mean that these are not in order? Can someone explain to me why they think that they aren't?" This questioning in response to actions on objects can prompt children to reflect on the idea of ordering.

Finally, encourage children to make a decision. "Of course you don't have to, but you may want to see if you can put your objects in order like Grant was doing last time. I'm going to be walking around, and I'd like to find out some of the ways that you order them."

The Activity Period

My wife and I have nicknamed our 2-year-old son Dr. No after his affinity for that word. But we could just as easily call him Activity-man. Like nearly all young children, Benji lives for activity, for moving and touching, building and demolishing.

Activity is at the heart of math and science workshops, in part because that is what children do best. The range of activities that children can do during the activity period was described in Chapter 4. This chapter addresses the teacher's role during all this activity. But first, the benefits of hands-on activities are expanded upon.

HANDS-ON ACTIVITIES AND COGNITIVE DEVELOPMENT

Hands-on math and science have several benefits for children. Aside from motivating children and the effects on math and science achievement documented in Chapter 2, experiences with physical objects are essential for intellectual development.

Cognitive Structures

A child's ability to think logically is dependent on mental tools called cognitive structures. According to Cohen (1984) "(Structures) are the mental mechanisms humans use to manipulate and act on data—that is, relate, interpret, synthesize, clarify, order, predict, infer, and hypothesize. Thus, one of the most important goals of all educators should be to facilitate the development of these structures" (p. 770).

Structures and Learning   Research has shown that a child with well-developed collections of cognitive structures tend to learn more than children with less-developed structures. Two structures that develop during the primary year—class inclusion and conservation of number—correlate to a wide range of mathematical abilities (see Hiebert and

Carpenter (1982) for a review). For example, first graders who possessed these structures were more likely to use advanced strategies such as counting on and derived facts to solve addition facts and performed better on all types of addition and subtraction fact problems (Hiebert, Carpenter, and Moser 1982). Another study found that students with a relatively large number of classification and spatial structures did better on science achievement tests, performing especially well on interpretive and evaluative questions that required abstract thought (Cohen 1992). Berg and Phillips (1994) concluded that certain spatial structures were likely necessary for children to construct and interpret line graphs.

Structures and Choice   What can teachers do to help their children grow cognitive structures? Since experiences with physical objects are needed to develop cognitively (Piaget 1964), teachers who use a lot of manipulatives are on the right track. But manipulatives are only part of the equation. Research shows that *how* children use manipulatives affects structure development. Students who were encouraged to follow their own interests while working with manipulatives—using, viewing, and holding them in different ways—showed far greater structure growth than children who used manipulatives only as directed by the teacher (Cohen 1983, 1984). Again, choice proves essential to learning.

Structures and Workshops

The workshop would seem to be perfectly tailored for children's cognitive growth. In four of the activity types—explorations, games, inquiries, and student-directed workshops—children consistently work with objects; during problem solving, object use is occasional. Additionally, nearly all workshop activities call for students to choose how to work hands-on. Teaching math and science through the workshop, therefore, would likely give children rich opportunities to aid their intellectual growth, which in turn would make math and science content more understandable.

Math Workshop Project Data   One purpose of the Math Workshop Project was to test out this idea. Would students taught workshop math and science have more advanced structure development, because they were frequently working hands-on and choosing what to do with objects? We looked at two cognitive structures: class inclusion and conservation of number.

Class Inclusion   If you ever try to play Twenty Questions with a child who cannot class include, you can see that he or she is at a serious disadvantage. Class inclusion deals with making relationships between groups and subgroups, an understanding that makes that game a lot easier. A child who could not class include would never ask umbrella questions such as, Is it a girl? Rather, because he or she cannot visualize the overall group and subgroups at the same time, this student will always ask about individuals: Is it Phillip? Is it Becky? Children who cannot class include have trouble understanding some math and science ideas, such as place value (10 is both 1 ten and 10 ones at the same time) and life science concepts (a flower is a part of a plant, not a separate entity).

Conservation of Number   Children who cannot conserve number believe that the quantity of objects can change when they are rearranged; their thinking is "ruled by perception rather than logic" (Phillips 1991, 257). One could imagine that it is extremely difficult to understand much about numbers if one believes that number is not stable.

Procedure and Results   This portion of the Project data was collected in a similar fashion to that described for number families in Chapter 5, using the same groups of students and teachers. The tests for the structures were interviews adapted from Phillips and Phillips (1996). Children were given points for each step of structure development they demonstrated, with a maximum of two points for class inclusion and three points for conservation of number. Figures 7–1 and 7–2 show the posttest results for the second-grade students.

GROUP	N	0	1	2	MEAN
Workshop	24	1 (4%)	11 (46%)	13 (54%)	1.48*
Non-workshop	20	7 (35%)	6 (30%)	7 (35%)	1.00

*Results were analyzed using analysis of variance (ANOVA). The workshop group mean was significantly higher at the $p < .05$ level.

FIGURE 7–1   Posttest results for class inclusion task.

GROUP	N	0	1	2	3	MEAN
Workshop	40	0 (0%)	14 (35%)	10 (25%)	16 (40%)	2.05*
Non-workshop	30	8 (27%)	14 (47%)	4 (14%)	4 (14%)	1.13

*Results were analyzed using analysis of variance (ANOVA). The workshop group mean was significantly higher at the $p < .0001$ level.

FIGURE 7–2   Posttest results for conservation of number task.

The data confirm that the workshop approach to math and science teaching has a very positive effect on cognitive development. Given the preceding research on the connection between development and learning, workshops appear to lay a firm foundation for math and science understanding.

THE TEACHER'S ROLE: CONFERENCING

In writing workshops, the teacher meets with individual students as the rest of the class writes. The young authors share their writing and thoughts in these student–teacher conferences. The teacher listens, asks questions, makes suggestions, and offers encouragement. Often the teacher takes notes such as, "Matthew used rich language in his writing; he should share this with the class." Conferencing is a private moment between adult and child, a powerful teaching and assessment technique rolled into one.

Children and teachers also conference during math and science workshops. This conferencing takes place in the activity period. While the class is conducting explorations or inquiries, solving problems, playing games, or working with objects during student-directed workshops, the teacher's primary role is to conference with individual children.

What Do Conferences Look Like?

What takes place in a conference depends both on the child's actions and your focus.

Add and Grab: Addition Strategies   After the minilesson, children gather their dice and chips, sit on the floor in pairs, and begin to play Roll and Grab. I give them a minute to get settled in and then begin making my rounds. While children are playing games, I can usually visit each group at least once.

Today I quietly sit next to the first group and start to watch them play. My goal is to look for the strategies that they are using as well as for openings to encourage better strategies. Often the question, How did you figure that out? helps bring out students' strategies.

When I notice that Alex is still using the count-all strategy, counting each dot on both dice to find the total, I want to see if he is ready to use the more advanced strategy of counting on. I start by saying, "It looks like you counted all the dots. You got it absolutely right, but is there an easier way to do it?" Alex looks puzzled, so I decide to make a suggestion, informally modeling the counting on process. I ask, "How many dots are on this dice?"

When he responded that there are five, I cover that dice and point to the other showing two dots. "Now, there are five here under my hand,

and one more would be? Here I pointed to one of the two dots on the visible dice. "Six," he replies. "And one more would be?" When he responded that there would be seven, I say, "You're right, it's seven just like you got before."

Alex's partner Grace jumps into the conversation. "I do that sometimes," she said. "I just say the bigger one, and then I count up on the smaller one. Like this: six, then seven, eight, nine—nine!" Partners often volunteer their methods, which is another reason why game mates should be children of similar abilities. The strategies of a partner far beyond or behind Alex would not benefit him as much.

It is hard to gauge what Alex thinks of this strategy, and I want to be sure not to push him into something that he does not understand. So, I wait for his next turn. He rolls a four and a five, then counts up from the smaller number: "*Four*, five, six, seven, eight, nine." This is often an intermediate strategy between counting all and counting up from the largest number.

"No!" Grace cries. She explains how it is better to start with the larger dice because then less counting is required.

But Alex seems satisfied with his progress. I am happy that he tried this new strategy, and although Grace's comments may lead him to consider her strategy eventually, I want him to experiment with his method for a while. So I say, "Do it however you think is best, Alex."

Add and Grab: Using Counters to Go Beyond Addition Strategies   Good activities offer a wide range of important ideas for children to learn. What children learn is, in part, determined by what you focus on during conferences. In Roll and Grab, for example, the counters that children win offer tremendous opportunities for learning beyond addition fact strategies, such as place value, conservation of number, and addition of larger numbers. Occasionally you can direct children's attention to the chips, especially with children who know their addition facts and who may need additional challenges.

During the game, children are interested in finding out who is ahead and by how much. To help keep track they often spontaneously group their chips in easy-to-count groups, often fives or tens. This situation is ripe for asking questions that may guide students toward place value understandings. For example, a boy who is grouping his by 10 volunteers that he has 45 chips. I can ask, "How many piles of 10 are in 45? How many are left over? When you say 45, show me the chips that are the '40' part of 45. Which chips are the '5' part?" Next I write '45' on a slip of paper and underline the four in the tens place. "Can you show me with your chips what this 4 means?" Then I underline the 5 ones. "Now show me with your chips what this 5 means."

Other children judge who is winning by looking at the piles and seeing which one looks bigger. Children who cannot yet conserve number often assume that the pile that is most spread out contains the most. By pushing the chips in the spread-out pile closer together and asking, "*Now* which pile has more?" you can prod nonconservers into rethinking their understanding. Those students with more advanced conservation abilities sometimes put their counters into rows, aligning their counters one-to-one with their partner's. Questions such as, "Who has more? How many more?" and "How do you know?" can promote further conservation development.

Children can also learn strategies for adding larger numbers as they count their chips. For example, consider a child who has 24 chips before her turn and then wins 8 more. She can either start counting her whole pile from the beginning or begin at 24 and count up 8. As with place-value concepts and number conservation, comments from a teacher can spur this advanced counting strategy. Simply by suggesting, "Now you have to start counting all over again. Remember that before you had 24. Could you just count 8 more to get your total?" you can lead a child toward this strategy.

Plant Inquiry   Because plants and animals do not handle touch well, the main activity-period activity for life science is observing and recording the organisms. During our plant inquiry, children sketched the coke, water, and coffee plants in their observation logs. Conferences during this time focused on both improving the quality of children's drawings and helping them think about their observations.

Conferences for Improving a Product   When the conference focus is a product, such as a drawing, piece of writing, or a model, it is helpful to include both a compliment and a suggestion. For example, "You really have the shape of the leaf perfect. Good job! Now take a look at the sizes of the plants in your drawings. You have them all the same height, but are the real plants all the same height?" A related strategy is to turn the evaluation over to the student: "What do you think is the best part of your drawing? Which part aren't you happy with?" These help the child to critically judge his or her own work. Chapter 9 describes how to encourage this self-assessment in detail as students observe tadpoles.

Conferences for Drawing Thoughts from Observations   While children observe and draw, conferences can also be used to help students think about the plants. A drawing of near-photo quality does not necessarily mean that that child has thought about his or her observations. Questions that focus children on the important ideas are part of productive conferences.

For example, I asked Guisti as he was sketching why he thought the water plant had grown the tallest. "Maybe because water has the most vitamins?" he replied. "Cola and coffee don't have many vitamins. They aren't very good for your body."

Neither water, cola, nor coffee have vitamins, of course, but Guisti's comment was grounded in truth: water is arguably the healthiest of these three drinks, at least for humans. So although it was important to set the record straight about vitamins, I wanted him to expand on his core idea. His extension beyond plants' environment to the environment of other living things fulfilled an objective of this unit. "Actually, water doesn't have any vitamins. Neither does cola nor coffee. But you are right that water is better for people than the other two drinks. Do plants and people need some of the same things?"

Guisti considered this. "Well, they both need air. And food. I guess plants and people do need some of the same things."

"Let's bring this up during reflection," I said, concluding the conference.

This 1-minute conversation accomplished several things. First, I was able to point out the misconception about vitamins. Although this was not relevant to the unit, it was easy enough to do. Second, Guisti had refined and rehearsed a hypothesis about plant growth.

Third, I saw that at least one student was able to extend his thoughts about environment beyond plants. It is usually best to wait for a student to initiate shifts in thinking such as this. If I were to suggest on my own that both plants and people depend on their environments, I could not know for sure if students were ready for this idea. That a student brought it up in the first place suggests that they may be able to handle this broader concept and that the time is right to present and debate it in the reflection period.

Sometimes students respond to questions such as, "Why do you think that the water plant has grown the tallest?" with silence or "I don't know." In these cases you may find it useful to repeat another child's idea. For example, "Lisa thought that maybe the water plant was closer to the window than the other two plants. What do you think about that?" Children often find it easier to consider the ideas of others than to make their own from scratch, and understanding can spring from thinking about other children's thoughts.

Searching for Seeds: Drawing Knowledge from Experience   The children were on a hunt, looking for seeds in our school garden. It was early fall, and unpicked squashes, apples, and beans were mixed among the dried seed heads of perennial flowers and grasses. Children excitedly shared their finds. They broke open shriveled green beans to show the pearly white seeds and demonstrated to their friends how to extract the seeds from marigold heads. Later, back in the classroom, the teacher asked, "So what part of the plant do seeds come from?" The same children who so intently collected seeds from fruits and dried flower heads just minutes before replied: "The roots!" "From inside the leaves!" "Seeds come from the stem!"

Students often ignore or explain away experiences, even hands-on experiences, with which they seem enthralled. This phenomenon is not limited to children. The history of science is full of incidents in which adult, professional scientists failed to notice relevant details that conflicted with their present beliefs (Chinn and Brewer 1993). Because children do not always perceive important details on their own, one role of the teacher during conferences is to point out what is important.

To illustrate, let's go back to the school garden as the children searched for seeds. A child comes up to you, proudly demonstrating how to tear a drying marigold flower open to collect the seeds inside. "That's a lot of seed," you say. "What part of the plant did they come from?" When the child replies that they came from the flower, you say, "Right! Can you find other flowers with seeds inside?" The child scurries off, now looking specifically for seeds in flowers.

You approach another child. He has squashed a crabapple and is examining the seeds. "What part of the plant did those seeds come from?"

"The fruit!"

"Evelyn already found seeds in a dry flower, and you found some in a fruit. Do you think that you can find seeds that are not in a flower or fruit?"

This conferencing strategy leads children toward the connection between plant parts and where seeds can be found. The challenge of finding seeds in any other part of the plant aside from the flower or fruit is especially powerful.

Pointing out the important aspects of activities during conferences helps students make the most of their experiences. Later, the reflection period is another time to drive these points home: "A lot of kids think that there are seeds in the roots, leaves, or the stem of the plant. From what you saw today, are they right? Did anyone find seeds in those plant parts? Of course not! Seeds are found only in seed holders such as dried flowers and fruits."

Student-Directed Workshop

Children are involved in an incredible variety of activity during student-directed workshops, so conferencing focus also needs to be varied. A good way to begin each conference is to say, "Tell me about what you are doing." This simple request gives children a chance to organize and communicate their thoughts. It also allows you to mentally plan a line of questioning that will build from that child's actions and thoughts. Figure 7–3 shows some questions and suggestions that you can make to children that connect to how they are working with their objects.

One Workshop   Following is an account of one workshop to provide a clearer picture of how to conference during student-directed workshops. By this time of year (early December), the 25 first graders are familiar with the rules and expectations of free workshops. At the end of the mini-lesson I called students two at a time to choose their objects, find their workspace, and to begin working. After waiting a minute to let the children get settled, I began my conferencing rounds.

STUDENT BEHAVIOR	QUESTIONS AND SUGGESTIONS
Sorting Children often organize their objects into groups that are similar in some way.	• What are the names of your groups? • Can you do it another way? • Can you put them into groups so that you have only (3) groups? • Which is your biggest/smallest group?
Counting Students are sometimes curious about how many are in their collections.	• How many do you have so far? • Can you estimate how many there are all together? • Is there a way that you can group them so that counting is easier? • You have three piles of ten and seven left over. How many all together? So if there are 37, can you show me the objects that are the "thirty" part of 37? How many piles of ten in 37? How many more would you need to have 40? 50? • How many objects are in these two piles together? Which one has more? How many more? • (With two piles of equal number): Are there more in this pile or that pile? Now watch as I spread this pile out, and push the other pile together. Which one has more now? How do you know?
Building up or out Many children like to stack blocks, or lay out other objects end-to-end.	• How many objects long/tall is that? • Is it as long/tall as you? How could you find out? • How many (Unifix cubes) long are you? • Can you see anything in this room that is as long/tall as your tower?

FIGURE 7–3   Student-directed workshop conferencing questions and suggestions. (Adapted from Phillips and Phillips 1996.)

STUDENT BEHAVIOR	QUESTIONS AND SUGGESTIONS
Building geometric designs Children often make two- or three-dimensional designs.	• Tell me about your design. • Can you think of something that is the same shape as what you built? • (If symmetrical) Pretend that your finger is a knife. Can you cut your design down the middle so that each side is the same? That's called the line of symmetry. • (If symmetrical) If I add a piece on this side, is it still symmetrical? Show me what you have to add to your side to keep it symmetrical.
Patterning Students frequently make linear or circular patterns with their objects.	• What is your pattern? • Close your eyes. I'm going to change around your pattern a little bit and when I'm done, open your eyes and tell me what I changed. • Can you build another design that is opposite of your first design?
Ordering Some children order their objects by length or overall size.	• What is your order? • Which one is the shortest? longest? • Close your eyes. I'm going to move them around, and you see if you can tell what I moved. • You ordered these by length. Can you order them in another way?

FIGURE 7–3   Continued.

MAX AND HIS BOTTLE CAPS

As I walked around the room, I noticed that Max was beginning to sort his collection of bottle caps into groups. "So, Max, what you are doing?"

"I grouped them, and then I'm going to use a calculator to add each group," he replied.

I wanted Max to verbalize his sorting strategy, so I asked, "How did you group them? What are the names of your groups?"

He explained that he had grouped them by the type of drink—Coke, Sprite, and Tab—that the bottle cap came from.

Rather than ask him other ways that the caps could be sorted, I decided that greater mathematical possibilities lay in his interest in counting them. He needed more time to put the caps into groups, so I told him that I would check back with him later.

PHILLIP AND HIS PATTERN BLOCKS

Phillip had the pattern blocks spread out in front of him when I arrived. I saw that he had covered a hexagon with a triangle, parallelogram, and rhombus (Figure 7–4). I recognized that by making congruent shapes (shapes that have exactly the same shape and size) such as this, Phillip could build important geometric and numerical relationships. "What did you do here?" I asked.

"I covered this shape with these other shapes."

"So you used this triangle, parallelogram, and rhombus to cover this hexagon? Interesting!" I replied, taking the opportunity to introduce some vocabulary. Then I issued a challenge. "Do you think that there are some other ways to cover a hexagon?"

Phillip started to experiment. After trying and rejecting several combinations, he discovered another congruent shape.

"Wow!" I exclaimed. "So two parallelograms also cover the hexagon! Do you think that there are any other ways? I'll come back in a few minutes to see what you found out."

LOIS AND HER UNIFIX CUBES

After observing several other children from a distance and deciding that they were making good progress on their own, I came to Lois. She was building with Unifix cubes on a tabletop.

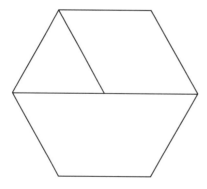

FIGURE 7–4   A child's geometric creation.

"What are you doing, Lois?" I asked.

"Just building a tower," she replied.

Since Lois was already thinking about height, I wanted to pose a measurement problem using Marilyn, Lois's partner, who was sitting nearby. "That's pretty tall," I said. "Do you think it's as tall as Marilyn?"

"It's almost as tall as me!" Marilyn replied, drawing an imaginary line in the air with her finger between the top of her head and the top of the tower.

Lois immediately saw something wrong with this: the tower was sitting on the table so it wasn't as tall as Marilyn, only as tall as the distance between the tabletop and Marilyn's head. "I'm doing it on the table, not the floor," she replied.

"What do you mean?" I asked.

"If you only measure it from the table, it's only as long as part of her," she reasoned.

Marilyn, clearly taken by this problem, suggested that she stand up and that Lois put the tower on the floor.

"It will fall down," Lois replied. "But I can make it on the ground and see if it's as long as you lying down."

I was satisfied that Lois and Marilyn were both interested and had a plan, so I left, promising to come back soon.

MICHAEL AND HIS COLOR TILES

Michael had sorted his tiles by color and was stacking each tile color into a separate tower. He replied to my opening question that he was trying to find out which tower would be the tallest. I was struck by the similarities between Michael's towers and the graphs that we had worked on the previous day. "Doesn't that remind you of a graph?" I asked.

"Yes! And so far the reds are the highest."

"Do you think that red will be the highest when you are all done?" I asked, wanting him to make a visual estimate of the remaining unstacked tiles to make a prediction.

Michael considered. "I think so," he said. "Either that or greens. There are a lot of reds and greens and not many of yellows and blues."

I thanked Michael, letting him complete his towers. "I'll be back soon to see if your prediction was correct!"

MAX AND HIS BOTTLE CAPS—PART TWO

When I returned to Max, he had completed his piles and was beginning to add up the subtotals on a calculator.

"How many so far?" I asked.

"Forty-one. I started with the lowest group."

I saw an opportunity to correct Max's terminology. "What do you mean 'the lowest group'?"

"I mean this one," he said, pointing to the Sprite pile. "It's the lowest."

"Oh! I thought that you meant that one of your piles was real low, like on the ground!" I joked. "So you started with the pile that had the fewest number? Which one did you do next?"

"This one. The pile that has a little more."

"So you sorted them into piles, ordered the piles by number, and now you are counting and adding. You've been very busy!" I said. I wanted to make him aware of the wide range of math that he was doing just by following his interests. Then, seeing the chance to sharpen his estimation skills, I asked, "If you have 41 so far, what is your estimate for the total, counting all of the piles?"

He studied the remaining piles. "Two hundred. These piles are bigger. I bet that there are another 40 in this one, and this one is twice as big. I'll say 180 or 200." I recorded his estimate on a small piece of paper ("My estimate: 180–200") and left him to do his work.

LOIS AND HER UNIFIX CUBES—PART 2

When I returned to Lois, she had completed her line of Unifix cubes so that it was the full length of Marilyn, who was lying on the floor.

"So, how many Unifix cubes long is she, Lois?" I asked, and the partners eagerly began counting.

MICHAEL AND HIS COLOR TILES—PART 2

Michael's color tile graph was complete. "So, is green or red the most?" I asked, reminding Michael of his prediction.

"It was a tie," he reported as he interpreted his three-dimensional graph. "There are 19 red and 19 green."

"How many reds and greens do you think there are all to-gether?" I asked. Michael had a good sense of addition and I wanted to see how he would approach a problem such as this.

"19 plus 19. That's hard," he said.

"Well," I suggested. "What if it was 20 red and 20 green? How much would it be then?"

Quickly, Michael replied, "Oh! 38. It would be 38. Because 20 and 20 is 40, so 19 and 19 would be 2 less. That's 38!"

PHILLIP AND HIS PATTERN BLOCKS—PART 2

When I returned to Phillip, the same two congruent shape combinations were sitting on his table, and he was using the pattern blocks to make a two-dimensional design. "Did you find any other ways to cover that hexagon?" I asked.

"No. I decided not to do it. I'm making a design now."

Although I was a little disappointed that Phillip had not continued his original line of inquiry, I understood the importance of children following their own interests. Still, I took a note to share this congruent hexagon problem with the rest of the class during the next student-directed workshop minilesson. It could easily capture the interest of some other students.

LOIS AND HER UNIFIX CUBES—PART 3

While I was gone, Lois and Marilyn had taken the measurement problem several steps further than my original. After Marilyn had been measured, they had called other children over. The pair organized their data into a table (Figure 7–5). Their excitement while sharing their information was tremendous. "We measured all these people with Unifix cubes! Maria was the same as Erin; they're both 55. Adie is the tallest. Let's measure you, Mr. Heuser!"

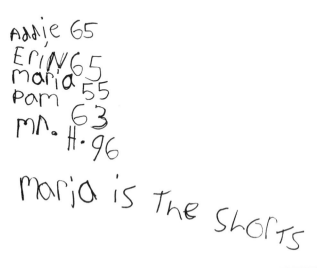

FIGURE 7–5   A child's recording of data that resulted from measuring people's heights with Unifix cubes.

CONCLUSION

Children capitalize on their interests, tackle difficult problems, and learn important math and science concepts in the activity periods of student-directed workshops. These accomplishments happen more often when teachers skillfully conference with their students. A successful conference begins and ends with the child. Teachers who find out what the child is doing and thinking can then use that knowledge to motivate and challenge.

Reflection Period

As a new teacher, I was always happy when my students were working with their hands. Touching was everything; if they were building, planting, drawing, sorting, or measuring, it was hands-on math and science, and that, I knew, was paramount. I rolled my eyes at colleagues who lectured their students while dust-covered manipulatives sat on shelves. Children learn hands-on, I was sure, not by listening to the teacher. Of course, in my inexperience I overlooked the same essential point as my lecture-giving colleagues: learning occurs not in the ears *or* the fingertips, but in the brain.

Reflection is the process in which the brain constructs knowledge as it organizes incoming stimuli. Whether the stimuli come out of the teacher's mouth or through a child's hands, learners need to reflect in order to construct knowledge. Without reflection, experiences will not be connected to past knowledge, and most of what is taught will simply be forgotten. For this reason, thoughtful and systematic reflection is an essential part of every workshop.

This chapter is about what teachers can do to promote reflection. In workshops, much of this reflection occurs in the reflection period that concludes every workshop. Keep in mind, however, that helping children to reflect is important throughout the entire workshop, and so some of these reflection strategies can also be used in the minilesson and activity period.

NURTURING REFLECTION

The best reflection occurs when teachers set the right classroom climate and when they plan and use effective reflection strategies.

Classroom Climate

What does a reflective classroom look like? There is much that you can do—that you are already probably doing—to cultivate a culture of reflection. Most of it occurs throughout the day, not just during math and science time. Do you stress and practice showing respect to everyone? Are children's ideas listened to, discussed, and valued? Is it acceptable for everyone—students and teachers—to make mistakes or to say I don't know? Nurturing these classroom characteristics takes time and effort, but they are essential for children to reflect.

PLANNING FOR REFLECTION

Aside from these foundations, there are specific teaching activities that can bring out the best reflection in your students.

Reflect on What?

Selection of reflection activities and prompts for each workshop is determined in part by what you want your children to reflect upon. Reflection in workshops has three purposes. Borasi and Rose (1989) defined several reasons for children to write journals in math class. I have adapted these for oral and written reflection in math and science. When children reflect, they can

1. build understanding of content and processes

2. improve their thinking strategies

3. express their feelings toward math and science topics

Figure 8–1 outlines prompts, benefits, and activities for each purpose. When planning a workshop, I decide what I would like children to reflect upon. Most often I use reflection to build understanding, but occasionally I want children to concentrate on their thinking or feelings.

Reflection Activities

The reflection activities shown in Figure 8–1 are usually used during the reflection period. The reflection period follows the activity period in workshops, though usually not immediately after. Often students need time to recharge before they are ready to reflect. Sometimes I will take them outside for a few minutes, or I may wait until after a special or lunch to begin the reflection period. It is wise not to put it off for too long—reflection should ideally happen on the same day that the activity period ends—because children (and teachers) soon begin to forget the fine details of their activities.

PURPOSE	PROMPTS	TEACHER BENEFITS	STUDENT BENEFITS	ACTIVITIES (ORAL AND *WRITTEN*)
Build Understanding	When would you use this outside of school? Tell someone what you know about . . . Explain what the 2 in 29 means. How has your tadpole changed? What questions do you have about . . . What did you learn? How could you solve this problem? What are the differences between a square and a rectangle? What are the four ways that they are alike? Name five ways that rock can be broken down. How does knowing that $51 - 30 = 21$ help you solve $51 - 32$? Write a story problem that you could use multiplication to solve. What would the world be like if there were no spheres? How are this length of wire and this screw alike? How are they different? How is the human life cycle similar to the frog life cycle? What's faster to make, a picture graph or a bar graph? What would be the best graph to show to a 4-year-old? Why? Football is one game that you use addition to keep track of the score. What's another one?	Provides topics for future activities, such as inquiries. Helps teachers reflect on and respond to their own teaching. Can be used as an assessment of children's understanding.	Helps students transform experience into knowledge. Helps students remember. Activates prior knowledge. Sharpens communication skills.	Pair/share* Conferencing* Class negotiation *Learning logs* *Observation logs**

FIGURE 8–1   Purposes for reflection.

PURPOSE	PROMPTS	TEACHER BENEFITS	STUDENT BENEFITS	ACTIVITIES (ORAL AND WRITTEN)
Improve Thinking	What's the easiest/hardest part of . . .? What problems did you encounter when you . . .? What mistake did you make most often? If you had to do your investigation again, how would you change it to make it better? How do you organize your work when you are solving a problem?	Provides topics for future activities, such as addressing common problems. Can be used as an assessment of children's inquiry and problem-solving strategies.	Gives insight into strengths and areas for growth. Communicates that problems are part of learning.	Pair/share* Class negotiation* *Learning log*
Express Feeling	How do you feel about . . .?	Provides topics for future activities that can address student feelings and attitudes.	Provides a therapeutic effect, as children realize that their teacher cares about their feelings.	Pair/share* Class negotiation* *Learning log*

*These activities can also be used in minilessons and activity periods.

FIGURE 8–1   Continued.

The reflection period usually includes both oral and written reflection. The research presented in Chapter 2 indicates that there are considerable benefits to both types of reflection. Moreover, having children reflect orally first and then write seems to especially promote learning.

Oral Reflection Activities

Class Negotiation   Class negotiation is the principal form of reflection in the workshop classroom. Negotiation is not a lecture. In Chapter 3 I presented a common scenario: a class discussion in which the teacher tried to teach what she felt was important about magnetism by telling students what they learned while they worked with magnets. Class negotiation is quite different than this.

Class negotiation is the process of consolidating individual ideas into shared beliefs. During each activity, children construct knowledge that is unique to each individual. Although this individual knowledge is essential, it is equally essential that the classroom community shapes learning. These shared understandings are facilitated in class negotiations as children present their thoughts, which in turn may be influenced by the thoughts of their peers and their teacher. Negotiation ends with the class either agreeing on a set of knowledge or agreeing to disagree.

Procedure

1. Determine the important concepts, facts, processes, or attitudes that are likely to be learned from an activity.

2. Begin the negotiation by stating a prompt that addresses one of these important ideas. The prompts in Figure 8–1 can be used as a guide.

3. Call on students to respond verbally to the prompt. Pair shares can also be used to help generate ideas prior to students presenting them to the whole class.

4. Ask students to react to the first set of student responses.

5. After students have had enough time to present, debate, and refine the first idea, write a brief statement on the board summarizing the group's conclusion.

6. Steps 2–5 can be repeated for other important ideas.

It is important to note that these steps, like the others in this chapter, are only a guide. The amount of time you have and the energy and interest levels of students will affect how much your negotiations follow the preceding steps. For example, sometimes negotiations last all of 2 minutes and we skip steps 5 and 6. At other times, students are still interested and learning after half an hour, and we fill the whole chart paper with our negotiated statements.

Examples

1. *Negotiating strategies for addition:* As students played Double Roll and Grab in previous workshops, I noticed that many children who seemed ready to use derived-fact strategies were still counting up. Consequently, I planned on having children debate the efficacy of both strategies during the next workshop reflection period game. I felt this debate would probably lead more children toward using the derived-facts strategy.

   As the children sat by the white board for the reflection period, I drew pairs of dice on the board: 9 and 8, 8 and 7, 5 and 7. I selected these pairs because each is close to a double and can be solved by deriving the answers from the doubles' facts. I then stated the prompt, "How would you figure out these sums if you rolled these dice in the game? Take a minute to solve as many of these as you can. Raise your hand when you have at least one answer. Then I'd like to find out your answers and how you got them."

When most children had their hands up, I began to call on them. William responded first. "I'll do the first one (9 + 8). It's 17."

"How did you solve it?"

William responded, "I just knew it. I have it memorized."

"Good," I replied. "Eventually we all want to just know all the addition facts. But let's pretend that you don't have it memorized yet. How else could you solve it?"

I called on Lisa, who I knew had counted up on a similar problem during the game. "I think it's 17. I started at 9, and then I counted up 8 more on my fingers. So that's 17."

I wrote 9 under that die and then under the 8 die, 10, 11, 12, 13, 14, 15, 16, *17* in order to show Lisa's strategy visually. Then I said, "OK. That's one way. Does anyone disagree with this answer or have another way to solve it?"

Another child suggested the use of tallies, so I followed her direction by tallying 9, then 8 more and counting all the tallies.

I then called on a child who frequently used derived facts. Tamika replied, "I knew that 9 plus 9 is 18. So since 8 is 1 less than 9, I knew that the answer would be one less also. So that's 17." As she spoke, I recorded her idea on the whiteboard (Figure 8–2).

Wanting to capitalize on Tamika's idea, I pushed for more details. "Why did you do it that way? Was there something about 9 and 8 that made it easier for you?"

"I know all of the doubles by heart. So it's just easier if it is close to a double," she replied.

"How many of you know the doubles by heart?" I asked. I wanted to be as explicit as possible that this may be a good strategy for some of the children, so when most of the children raised their hand, I asked, "How many of you use what you know about doubles to solve ones that are close to doubles? That can be a great strategy. If this makes sense to you, try it on the next pairs."

Other children then solved the problems and explained their strategies. Several applied derived fact strategies, and others used other methods. After taking several more explanations, I asked a question that I hoped would prompt those children who were ready to adapt strategies using derived facts. "All these problems are similar in a way. What's the best way to solve problems like these?"

Tamika knows that $9 + 9 = 18$

8 is one less than 9,
so the answer will
be one less: $9 + 8 = 17$

FIGURE 8–2   Tamika's solution strategy for solving $9 + 8$.

"Any strategy is good, if it makes sense to you," volunteered one child, echoing my refrain that students should use problem-solving methods that they understand.

Although I was happy that the children believed this, I wanted to impress on them that some ways are better than others. "But what if you understand more than one strategy? Like what if you get the counting up idea *and* using the close-to-doubles idea. Which one should you use then?"

"Using doubles facts is better," responded Eric. "It's faster."

Another child chimed in. "Plus, you'll make less mistakes. Say you were counting up 8 and you miscount." She mimicked miscounting on her fingers: "14, 15, 16. So you got the wrong answer."

I replied, "I see what you mean. It's faster and you're less likely to make mistakes by using facts that you know, like the doubles. So if this makes sense to you, give it a try! What if I write this: 'If a problem is close to a doubles fact, it can be good to use the doubles fact to solve the problem.' "

"Why don't you just say *always* use the doubles fact? It's really easy," said Tamika.

Robert responded, "Don't put that! I like to use another way."

"Robert has a good point." I said. "Not everyone uses doubles facts. Shall we leave it like that then?" I asked, and when there was no objection: "OK, I'll write that."

2. *Negotiating the relative benefits of coffee, water, and coke:* Two weeks into our plant inquiry, the coffee and the water plants were jockeying back and forth for the lead, with cola far behind. I posed this prompt to the children to help them construct some understandings from these events: "What is better for a plant: water, coffee, or coke?"

Students immediately responded that cola was definitely *not* the best, because that plant was "in last place." I wanted to get children to expand on this, beyond simple observation of the plants' lengths. "I see your point," I said. "The cola plant is definitely the shortest. Why do you think the cola plant isn't doing as well?"

One child shared her idea. "Maybe the coffee plant didn't get as much sun as the others did." She noted that it looked like that plant was farther away from the window than the other two. I was happy that she was considering variables other than with what the plants were "watered," but other children quickly jumped in. "No! Remember we put them all the same from the window so that it would be fair." This gave me a good chance to remind students how we tried to control variables. "Remember that we tried to make it as fair as possible, so we used a ruler to make sure the plants would all be the same distance from the window?"

Another child offered, "Maybe cola is bad for plants, just like it's bad for kids." Others agreed, sharing how their parents had cautioned them about drinking too much pop. It was significant that they were trying to draw connections between different types of living things. Still, I did not want them to reach faulty conclusions. "I think that we can agree that, so far, coffee and water seem to be better than cola for plants, but I wonder if that is the reason. After all, coffee isn't as good as water for people to drink, but the coffee plant is the longest." So for the first part of our negotiation, we wrote: "So far, cola is the worst for plants. Maybe it is bad for plants just like it is for people. We should look it up in a book to find out for sure."

Next, I asked why the water and coffee plants seemed to be growing so well. Ideas were presented and responded to: "Coffee has caffeine, and caffeine makes things go fast," was one idea. "True," I said. "But cola also has caffeine, and that one is doing the worst." "Maybe the winning plants got more to drink than the coke?" was the next idea. "No!" came the response from several children, and they recounted how we had planned to give each plant exactly one half cup of liquid every other day.

Still other thoughts were presented. One child volunteered that his grandmother put coffee grounds in her compost pile, "so coffee is probably good for plants." Several children suggested that since it rained water, plants would already be accustomed to water.

At this point I wanted children to generalize about how the plant's environment affects its growth and health. "It seems like we have a lot of ideas, but we are not sure *why* coffee and water are doing better than coke. But, do you agree that what you give a plant can affect how it grows?"

"Yes," came the reply, and when I asked for an explanation: "The coffee plant was always slower than the other two," and "If you feed a plant poison you know it will die."

Finally, wanting children to extend this idea beyond the liquid a plant receives to how environment affects growth, I posed a final prompt: "What other things besides what a plant drinks affects its health?" Children quickly supplied other environmental factors, such as amount of light, quality of soil, air temperature, and degree of "respect:" ("If people or animals pick at plants they might die.") I summed up this discussion by writing, with the students' help: "There are a lot of things that help (or hurt) a plant in their environment."

This negotiation worked toward many of the goals for the unit outlined in Chapter 4:

- understanding that environment affects an organism's growth and health.
- learning inquiry skills, including conducting a simple investigation, employing simple equipment and tools to gather data and extend the senses, using data to construct a reasonable explanation, and communicating investigations and explanations.
- knowing that scientists review and question each other's work.
- knowing that scientists develop explanations using observations and scientific knowledge.

By having several similar negotiations during this unit, students were likely to develop a strong understanding of these concepts. Results of assessments shown in Chapter 9 illustrate just how successful the students were.

3. *Disappearing water: An extended negotiation:* Negotiations sometimes take place over several days, as shown in the following example. One year my class of first graders observed a classic water demonstration. We poured water into an open, transparent container, marked the water level on the container, and placed it on a sunny windowsill.

During the reflection period the next day, children noticed that the water level was now lower than it was originally. "Where did the missing water go?" I asked. "My mom told me that water goes up into the air," one boy said. "It's called evaporation." To some of the other students, this idea seemed laughable—how could water just go up into the air? Instead, they suggested some alternative theories: (1) the water disappeared, (2) the water leaked out a tiny hole in the cup, (3) someone accidentally spilled it, or (4) someone intentionally took some out (here they looked at me).

Rather than impose on each child the scientifically accepted truth, we began to negotiate some shared beliefs. I began this process by asking each group to explain why they thought what they thought. Some children began to form explanations based on past experiences. A boy shared that he had to occasionally add water to his fish tank at home. Debate ensued: "Maybe you have a leak in your tank," one of the children who held the hole-in-the-cup theory suggested. "I don't think so," the first child responded, "because the floor is never wet."

After some more debate, I realized that we were not going to agree on anything too substantial. I saw an opportunity to use these different interpretations to get more out of the water demonstration, so I asked how we could prove any of our ideas. Students had several suggestions, such as covering the top of the container with plastic wrap to stop any water from going into the air, and putting a paper towel under the cup to show any leaks. One child also had the idea that *he* should take the container home, so that I couldn't take out any water. We agreed that *everyone* should try it at home—I would provide them with marked plastic cups—and that we would report back in the morning.

Before we left, however, I asked children to tell me what we already knew. It was important that we summarized our beliefs to this point, even though they would likely change as we gathered more information. Eventually we agreed on this statement, which I wrote on chart paper: "Some water was missing from the cup. It might have disappeared, or maybe the cup had a leak, or maybe someone spilled the water. To find out, we are all going to do an experiment tonight."

That next morning students began talking about their experiments as soon as they came in the room. We postponed our usual morning meeting in order to capitalize on the excitement. The students who covered their cups with plastic wrap reported that their water levels hadn't changed. Some even noticed that some water had collected on the plastic wrap. "How did it get on the plastic wrap?" I asked. One child said that it "floated up there," but most proposed alternative suggestions, like the water sloshed onto the wrap when the cup was moved.

The paper-towel group shared next. All these cups had lost some water, and the paper-towels didn't look wet. So, most agreed, that the water probably didn't leak out. "So where did it go?" I asked the class.

"Water can just disappear," was the dominant answer. Even those children who used the term *evaporation* seemed, upon further questioning, really to mean that the water ceased to exist instead of being converted to an invisible form. So we wrote on a new piece of chart paper: "Water disappears if it is not covered up, but if you cover it, it won't disappear."

I think that these extra negotiations were worth the time. Rather than stick with our original statement that didn't really say anything, like real scientists we gathered more information and negotiated a more sophisticated conclusion. Even though this new conclusion appears very simple, it is consistent with research showing that since primary children are not likely to understand that water can be in the form of a gas, they believe that it simply vanishes (see Project 2061, American Association for the Advancement of Science (1993) for a review).

Pair Share   Pair shares are when two people share their ideas on a given topic. These topics are given by a prompt from the teacher. This is a powerful reflection strategy because every child in the class has to form and articulate his or her thoughts—no one is allowed to be passive. Pair shares are most often used to have students respond to a part of an activity or to a comment, as a way to activate prior knowledge, and, initially, to get children's attention.

Ideally, each pair would take turns responding to the prompt, carefully listening to their partners and thoughtfully replying, perhaps asking for clarification or expressing their agreement or disagreement. This ideal, of course, is not realistic for all children every time. That is OK; the main purpose of pair shares is to allow children to verbalize their thoughts. If their partner responds or even listens, it is an added bonus that can come eventually as children move into the later primary years.

Procedure

1. Say, "Turn to your neighbor and tell them. . . ."

2. When you sense that each child in most pairs has had a chance to share (most pair shares last less than a minute), say, "Three, two, one." This is the cue that students need to finish their thoughts and return their focus to the teacher.

3. Ask several students to share their comments with the whole class.

Suggestions

This strategy, like all others, has to be directly taught and practiced. Some children are hesitant to approach others to be their partners. You may have to direct those children to find another unclaimed person who is nearby or, still easier, have them join a pair adjacent to them to make a triad. During subsequent pair shares you can say, "Turn to *another* neighbor and share. . . ." This also encourages students to actively seek pair-share partners.

Make it clear that each child is responsible for expressing his or her ideas—no one can sit back and let the partner do all the talking. Likewise, some children may take up the whole time talking. If this is an issue you can give them a halftime warning, indicating that if the other partners haven't shared yet that they need to start.

Examples   Pair shares are a good way to help children reflect on and communicate what they know. For example, after writing several addition facts up on the board, I announced, "Figure out as many of these as you can. When you have a couple, turn to a neighbor and tell them how you solved one of them." This helped children reflect on their own strategies and consider others. Also, the students had already rehearsed communicating their explanations, which facilitated the group discussion.

Pair shares can also be used as a way to respond to comments. As we reflected on a rock-sorting activity, I said, "I heard one student say that you can't break rocks because they are too hard. Turn to a neighbor and tell them if you think that is true or not."

Activating children's prior knowledge of a topic is vital to involving them in minilessons. Pair shares can help students remember related experiences. For example, as you begin to study money, you could start a minilesson this way: "Do you have a piggybank or something else that holds money at home? Tell a neighbor how much you think you have."

Written Reflection Activities

Reflecting on math and science in writing has several advantages, not the least of which is that students are *writing*. In most workshops, students are writing five to ten minutes. This can help ease time constraints for teachers who feel that they have to schedule sufficient language arts time first, with whatever is left over to be used for math and science.

Kindergarten and first-grade teachers are not concerned so much about a lack of writing time but rather that their students cannot (or will not) write. Written reflection in math and science workshops does not have to wait until students can write perfectly. Figure 8–9 shows that like their writing outside of math and science, children's first reflective writings will be developmentally appropriate: often messy and unintelligible.

The same good strategies that you use to teach personal writing can be used to teach reflective writing. The teacher or other adult can transcribe children's writing. Drawings are an acceptable way to communicate important ideas. Guided writing, in which the teacher writes student ideas on an overhead transparency or chart paper, is another way to nurture math and science reflection and at the same time model the writing process.

Two written reflection activities are learning logs and observational journals.

Learning Log Learning logs are journals of students' math and science learning. These logs are often the final step of each workshop, and log writing encourages students to form and state personal connections between workshop activities, discussions, and past knowledge. Because log entries chronicle individual beliefs, responses will vary widely from child to child.

Students write in response to the teacher's prompts. Therefore, teachers need to select prompts carefully to match their intended purpose. Figure 8–1 outlines the three purposes of reflection as well as prompts matching each purpose.

Procedure

1. Write the prompt on the board or on an overhead, and read it several times to the class.

2. Set the time in which students are expected to write, such as, "You have 5 minutes to write as much as you can to this prompt."

3. Circulate and silently read some of the logs as they are being written, making suggestions and asking for clarification as necessary.

4. Collect the logs so you can read them later.

Suggestions

Children need considerable guidance as to what makes a good learning log entry. Quite a few minilessons and reflection periods are devoted to learning about this kind of writing.

Learning logs do not have to be in the form of a bound booklet. Although having an organized, sequential record of children's responses has its advantages, students can lose them, and the larger ones become difficult for the teacher to collect. One of the best spots for learning-log writing is on the back of activity period response sheets or on a single piece of lined paper.

The purpose of setting a time duration for writing is to communicate to children that writing about math and science is important, and this is the time to do it. Some children have the tendency to want to rush to the next thing. The time allotment says that you will not allow this important work to be rushed. If students do finish early, I say, "Well, you still have 2 minutes left. Try to write some more or quietly read it to yourself." Challenging students to fill a whole page is one way to thwart the "I'm done!" responses.

It is important to read student logs for two reasons. First, they usually give a pretty accurate picture of the success or lack of success of the workshop. Second, student-learning log reflections are essential for thoughtfully planning subsequent workshops.

While reading student entries is important, responding to students individually is less so. Because it is time consuming, I write back to individual students once every two or three log entries. I try especially to respond to any expression of feeling, excellent work, or clearly inaccurate thinking. Figure 8–5 illustrates one teacher response.

It is difficult for children to reflect on math and science, and at the same time pay attention to writing conventions. For this reason, it is all right for learning logs to be filled with inventive spellings and unconventional punctuation and capitalization.

Math Examples

1. *Encouraging derived facts:* To follow up on the class negotiation to encourage the use of derived facts, children wrote three ways to solve the problem 9 + 8. I chose this fact because it is fairly difficult to solve by counting up, but it can be solved by using doubles (8 + 8 + 1 or 9 + 9 − 1) and 10 ((9 + 1) + (8 − 1)). Figure 8–3 shows several responses. Note that (b) suggests, "Ask your teacher for a math game," reflecting the awareness that her teacher values games over worksheets.

2. *Applying knowledge of area:* In order to build a strong understanding of measuring area, students were given two different-sized paper rectangles and asked to find out which was bigger. Students invented several ways of finding out which had the greater area. Two common approaches were to cut one of the sheets into pieces and then to overlay the pieces onto the other shape and to cover each of the rectangles with like-sized objects, such as base 10 blocks or inch cubes.

   In class negotiations students presented and debated their strategies and then wrote in their learning logs the steps that someone could follow to calculate area. This last step helps students think through their invented procedure. It is also important that children be able to apply what they understand to solve new problems, especially those outside of school. So after writing out the steps, I asked them to respond to this prompt: "While you are talking to your friend on the phone, he says that his bed is bigger than yours, but you think that yours is bigger. How could you use what you know about measuring area to find out who is right?" This prompt was a challenge to students to adapt a strategy for a new environment—in bedrooms without math manipulatives.

   Responses showed several levels of success (Figure 8–4). Some students suggested using a ruler or shoes to measure how *long* each

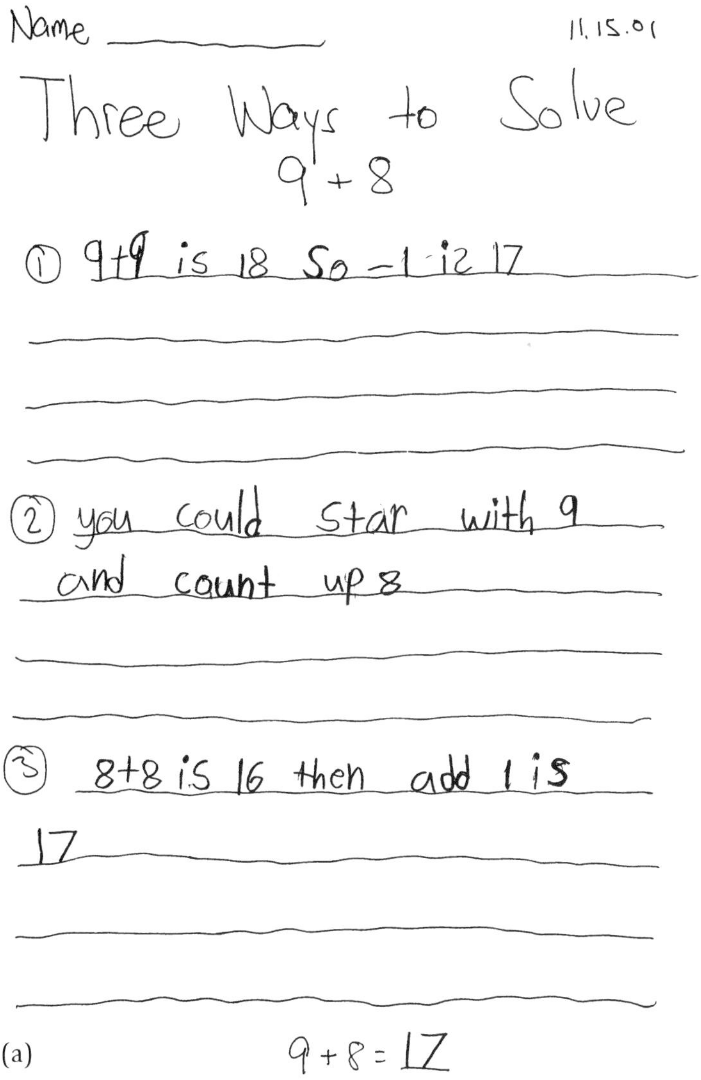

Name _____   11.15.01

Three Ways to Solve
9 + 8

① 9+9 is 18 So −1 i2 17 _____

_____

_____

_____

② you could star with 9 _____
and count up 8 _____

_____

_____

③ 8+8 is 16 then add 1 is _____

17 _____

_____

_____

(a)   9 + 8 = 17

FIGURE 8–3   Learning log responses to "Explain three ways to solve 9 + 8."

bed is but did not mention the width. Other students proposed
covering the beds with objects such as dogs or people (8–4(a)),
reflecting that area measurement has to be two-dimensional. One
student even suggested that looking at the label on the bed box
might help! Other ideas showed that some children had a very
advanced view of area. The child who wrote (b) not only carefully

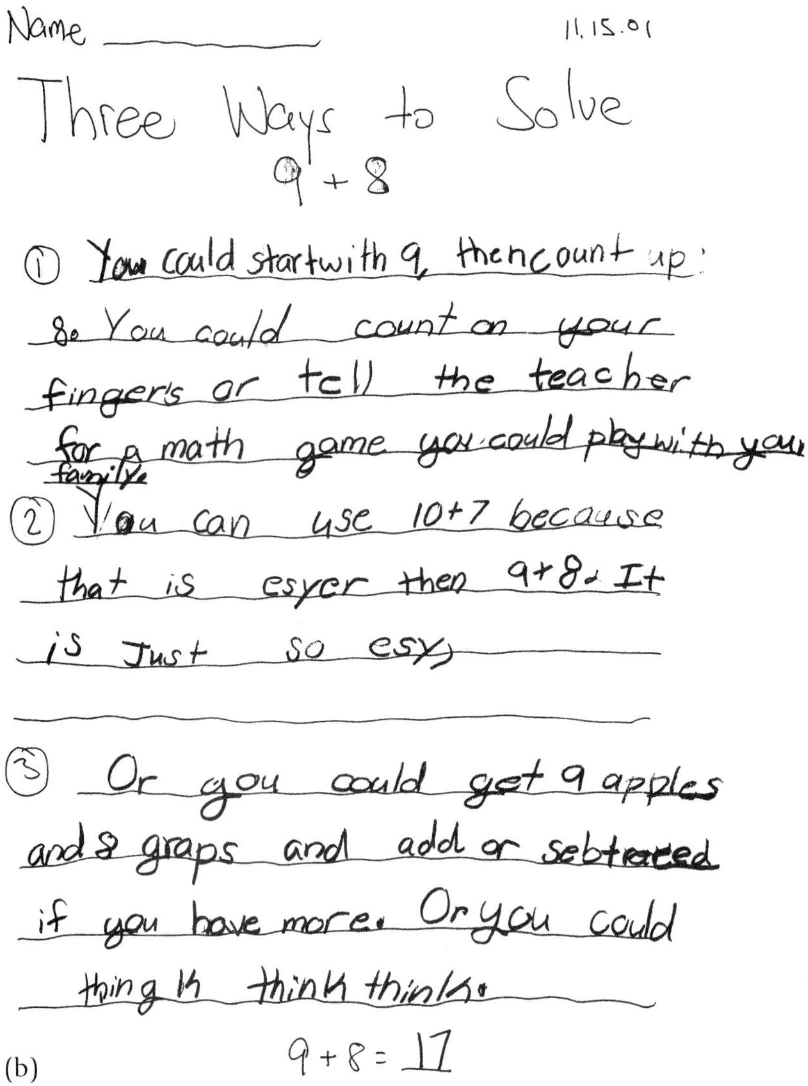

Name _____   11.15.01

Three Ways to Solve
9 + 8

① You could start with 9, then count up: 8. You could count on your fingers or tell the teacher for a math game you could play with your family.

② You can use 10+7 because that is esyer then 9+8. It is Just so esy.

③ Or you could get 9 apples and 8 graps and add or sebtrered if you have more. Or you could thing th think thinks.

(b)   9 + 8 = 17

FIGURE 8–3   Continued.

---

organized her response, but she also acknowledged that the process would result in a grid (step 1) and that the units had to be the same size (step 3). Learning log (c) indicates a child who is close to discovering the standard algorithm for area (length times width) on her own.

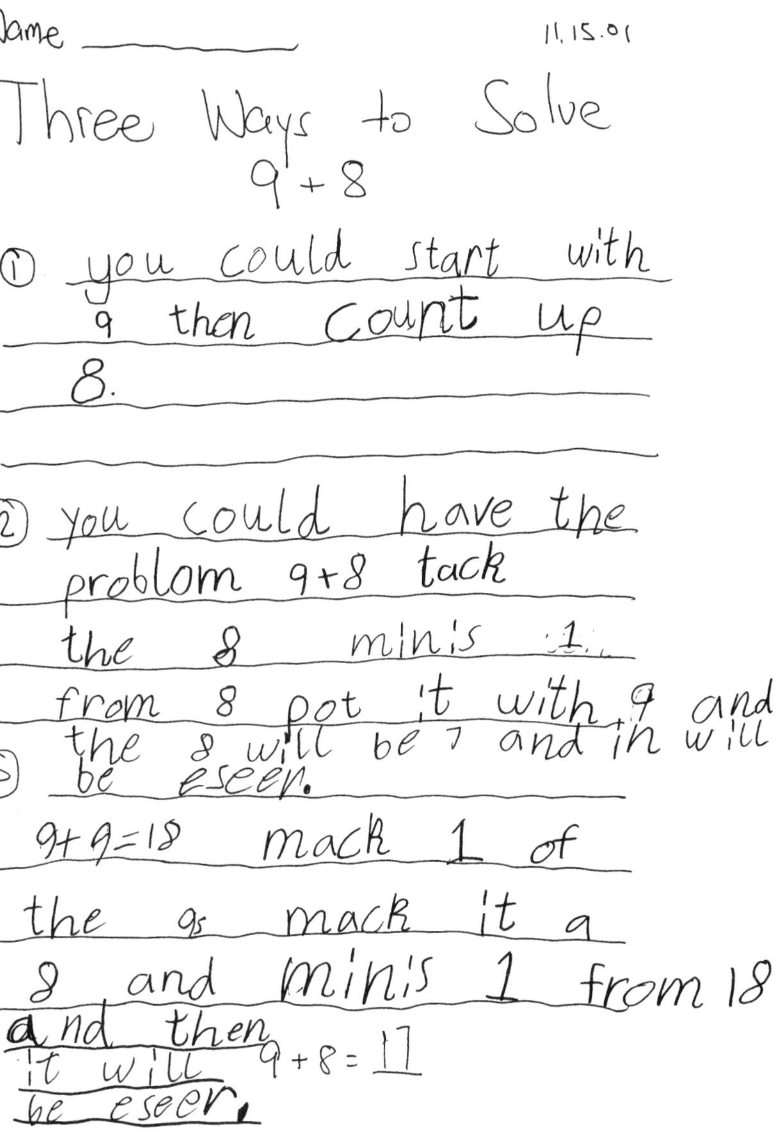

Name _____   11.15.01

Three Ways to Solve
9 + 8

① you could start with
   9 then count up
   8.

② you could have the
   problom 9+8 tack
   the 8 minis 1.
   from 8 pot it with 9 and
   the 8 will be 7 and ih will
③ be eseen.
   9+9=18 mack 1 of
   the 9s mack it a
   8 and minis 1 from 18
   and then
   it will 9 + 8 = 17
(c) be eseer.

FIGURE 8–3 Continued.

3. *Feelings about addition and subtraction:* One year I sensed that my
   students were feeling frustrated about multidigit addition and subtraction. I wanted to address any negative feelings, so I asked them
   to respond to some of these prompts in their learning log: How are
   you feeling about multidigit addition and subtraction? What is the

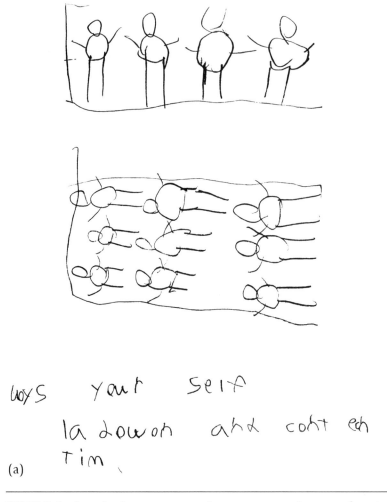

wosS yaur seif
la douon ahd coht ech
Tim.

(a)

FIGURE 8–4   Learning log responses on how to compare the areas of two beds.

hardest part for you? What can I do to help you add and subtract better? After reading their responses, I wrote back to the students to encourage them. An example is shown in Figure 8–5.

Science Examples

1. *Articulating Water Level:* An important part of physical science is understanding the characteristics of everyday materials, such as water. During our unit on solids and liquids, it became clear that students did not realize that the surface line of still liquids is always

Say Good-Bye to
your friend and before you hang
up with your friend say take
all the pillows in your house and
1. Put them in rows all
over your bed across.
2. count the pillows

3! make
shur your pillow and your
friends pillow is the same
size.
4. call your friend and
tell her/he How Many ~~money~~
pillows you put on
your bed in rows
(b) ACROSS

FIGURE 8–4   Continued.

level. For example, some children believed that water level would rise only on the side of a container into which water is poured or that liquid levels could remain parallel to a container, as in the first drawing in Figure 8–6. With this in mind, I tried to help students gain this awareness by challenging them over several workshops to show me a liquid among those on which they were working whose surface was not level.

At the end of one workshop in which we discussed this characteristic of liquids, each student wrote on the learning-log sheet

you could
take a rulr and
maser how
long her bed
is. Then
maser how long
your bed is
maser
acoss
and
donw

(c)

FIGURE 8–4   Continued.

shown in Figure 8–6. Nearly all correctly circled the first drawing as
not making sense. In order to respond to "Explain why those don't
make sense" children had to use their understanding about matter
in a new situation. Students did this in several ways.

The response in (a) shows that this child was able to apply
what she knew about water to milk. Other children drew relationships to other parts of our unit, such as the nature of solids (b) and
to viscous (thick) liquids (c).

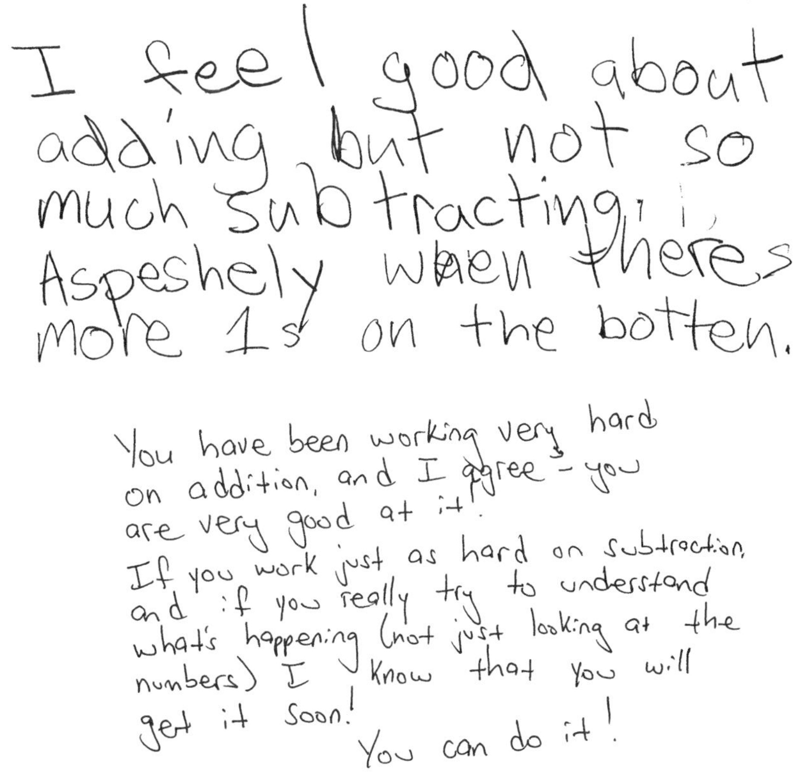

I feel good about adding but not so much subtracting, Aspeshely when theres more 1s on the botten.

You have been working very hard on addition, and I agree – you are very good at it.
If you work just as hard on subtraction, and if you really try to understand what's happening (not just looking at the numbers) I know that you will get it soon!
You can do it!

FIGURE 8–5   Learning log entry showing student's feelings and teacher's response.

Sometimes children surprise you with the sophistication of their explanations. One child, for example, didn't write directly about the milk but instead drew on what she understood about the nature of air (d).

2. *People and environment:* The environment affects humans, as well as plants. Following the workshop described before, in which we negotiated that plants are affected by their environment, it was important that students see that what is true of plants is also true of other organisms. So we went to the library to read books about what people need to stay healthy. Then, students drew and wrote about how environment affects human health (Figure 8–7). These learning logs—really learning posters—illustrate how important drawings are to processing and communicating thought. This is especially true of emerging writers.

Look at the glasses of milk. Circle the ones that don't make sense.

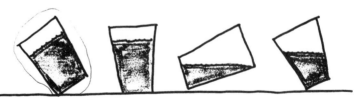

Explain why those don't make sense. Because I test it out on water.

Name _____

FIGURE 8–6 Learning log entry explaining why the first drawing is illogical.

Look at the glasses of milk. Circle the ones that don't make sense.

Explain why those don't make, sense. Because it has to be a straight line no matter which way you tilt it. It would have to be frozen solid.

Name _____

(b)

FIGURE 8–6 Continued.

Look at the glasses of milk. Circle the ones that don't make sense.

Explain why those don't make sense. Circeld glass car
exist   because only viscous liquids like Jello can' stay
        for a while.

Name _____
(c)

FIGURE 8–6   Continued.

Look at the glasses of milk. Circle the ones that don't make sense.

Explain why those don't make sense because the
milk line has to be a horizontal line
becase air floats to the highest
place it can get to, so the ai
would not be lower then the Mil

Name _____
(d)   129

FIGURE 8–6   Continued.

130

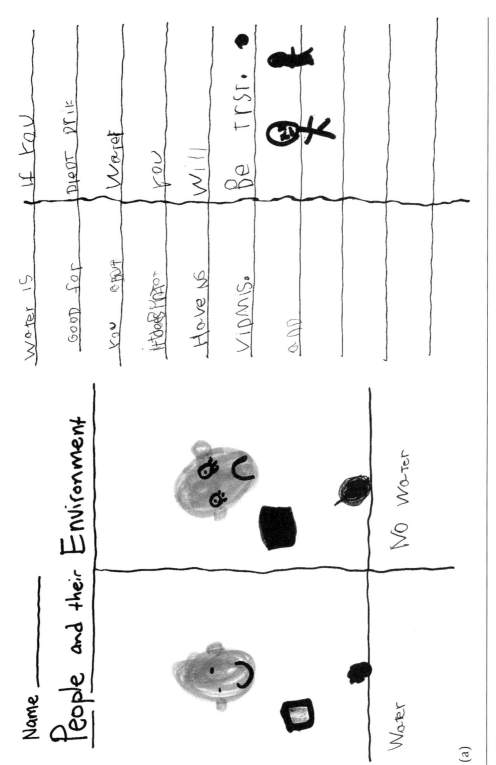

FIGURE 8–7   Learning log posters planning a plant experiment.

(a)

Name _____

People and their Environment

If I had If I had
air, I could no air, I
smell the would be
air and It's dede and
cood feel you can't
so nice. do enything
and I could you can't
play around do enything
a lot. when your
dede.

I did

(b)

FIGURE 8–7 Continued.

132

People and their Environment

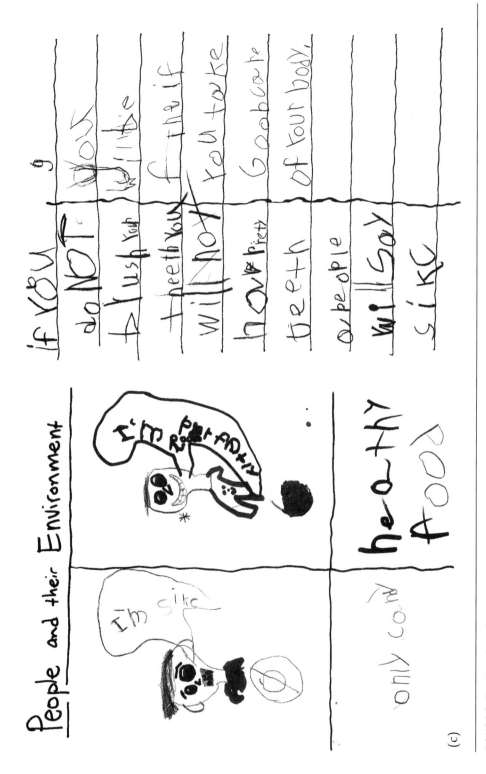

If you do NOT brush your teeth your teeth will NOT have Pretty teeth or people will say sike

I'm rotten parad+ty

I'm sike

healthy teeth only could Poop

3. *Flower to fruit: Birth announcements:* Some written responses can be short. To help children think about how the tiny pea pod emerged from the flower, I told students how new parents send out birth announcements and asked them to make their own cards that they could bring home to announce this blessed event. Children were tickled with this assignment and made very memorable announcements (Figure 8–8).

Observational Journals   Children write in learning logs after they have done activities, but observational journals are completed *during* the activity. These are most often used in science, to record children's observa-

FIGURE 8–8   "Birth announcements" as our pea plant bloomed.

tions of objects or experiments. Although they are not reflection period activities, writing and drawing in observational journals can promote a great deal of reflection. An example of how observational journals can be used for tadpoles is given in Chapter 9.

REFLECTION DURING STUDENT-DIRECTED WORKSHOPS

Reflection during student-directed workshops is so different than during teacher-directed workshops that it needs to be addressed separately. The reflection period of a student-directed workshop can consist of up to three activities: a pair share (always), a class negotiation (rarely), and learning-log writing (occasionally). Just what these activities look like in this setting is explained in the next sections.

Pair Share

Each student-directed workshop reflection period begins with a pair share. Pair shares give each child an opportunity to verbalize what he or she did and thus help contribute to vocabulary and conceptual development.

Getting students to successfully pair share takes a considerable amount of modeling and patience because (surprise!) little kids do not always listen very well to other little kids, especially when the teacher is not right there. Keeping it very structured during the first workshops helps. For example, after I ring the stop-working bell, I say, "Partner number 1, sit right next to your partner. Partner number 2, start to share your work." Then after about 1 minute, I ring the bell again and announce the opposite: "Partner number 2, sit right next to your partner. Partner number 1, start to share your work." Later in the year you can sometimes be less direct by simply ringing the bell and announcing that it is time to share.

Negotiation

Because students choose their own activities during student-directed workshops (and are thus working toward different understandings), negotiations occur only when a purpose of reflection is to improve thinking or express feelings. For example, you might start a conversation about the importance of pair sharing when you notice that some students are not sharing or listening well. Or, if a student complains that other students are intruding on her workspace, you could begin negotiations with the prompts, "How do you feel when someone moves into your space? What can we do about this since people want to have their own space to work in?"

Learning Log

At the very end of student-directed workshops, I frequently ask students to respond to a prompt by writing on the back of their recording sheet. I say "frequently" because we skip this writing if we run out of time or if students seem too tired. Given the diverse behaviors of students during these activity periods, simple prompts such as, "What did you do during the activity period?" or "What did you learn from working with your objects?" are usually best. Some samples of responses are shown in Figure 8–9 (a–c). Descriptive writing doesn't come naturally. Some workshop minilessons need to be devoted to modeling and discussing this type of writing to help students write effectively. Fortunately, this instruction also works toward meeting language arts goals.

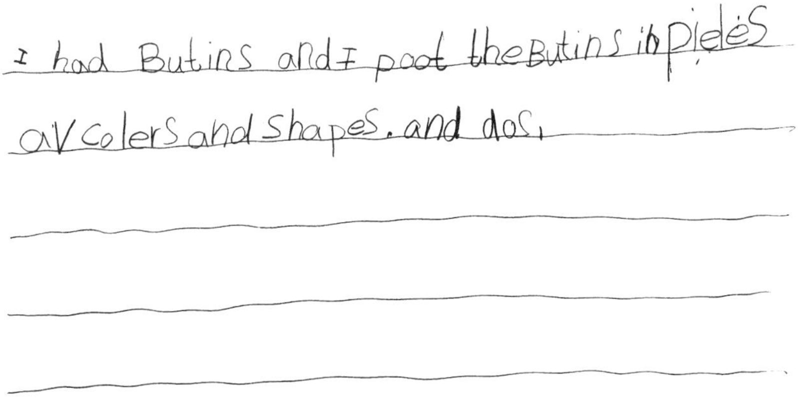

I had buttons and I put the buttons in piles of colors and shapes and dots.
(a)

FIGURE 8–9   Learning log entries: "What did you do with your objects?"

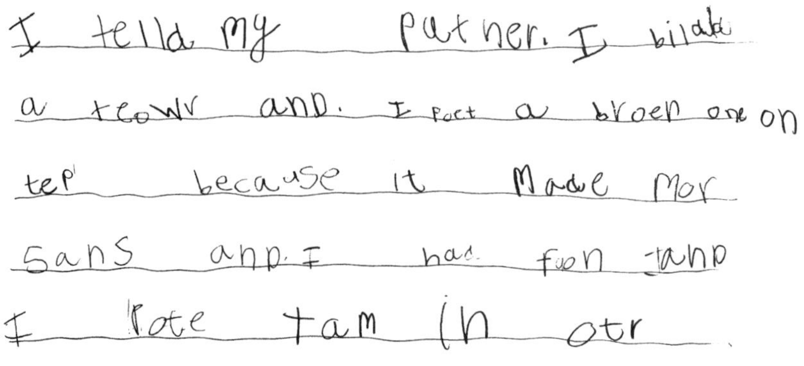

I tella my putner. I bilat
a teowr anD. I pot a broep ore on
tep because It Made Mor
5ans anp I had fon ianp
I lote tam in otr

I told my partner I built a tower and I put a brown one on top because it made
more sense and I had fun. I put them in order.
(b)

FIGURE 8–9   Continued.

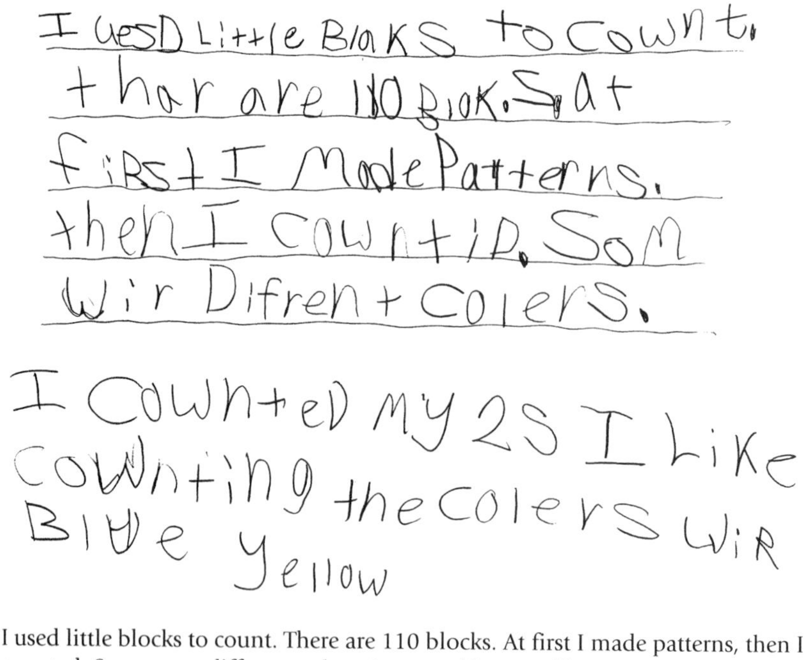

I uesD Little BlaKS to cownt.
thar are 110 BloK. Sp at
first I Made Patterns.
then I cowntiD. SoM
wir Difrent colers.

I cownted my 2S I Like
cownting the colers wir
Blue Yellow

I used little blocks to count. There are 110 blocks. At first I made patterns, then I
counted. Some were different colors. I counted by 2s. I like counting. The colors
were blue and yellow.
(c)

FIGURE 8–9   Continued.

9

Assessment

Workshops offer powerful opportunities for children to learn. But how do you know to what extent each child took advantage of these opportunities? What feedback during instruction may help children learn more? Even before instruction begins, what does each child know, what are his misconceptions, and is she developmentally ready to learn what you are teaching? Assessment helps to answer these questions.

FORMATIVE VERSUS SUMMATIVE ASSESSMENT

When I went to grade school, I took a lot of summative assessments. Of course, I didn't know them by that name. The names I knew were *pop quiz*, which were often given to see who paid attention during the lesson or who did the assigned reading. I also knew *chapter test*. These happened when my teacher got to the end of a chapter in her teacher's manual. In the past, summative assessments were often the only assessments given. Results from these tests were then used to compare children and to provide grades to go onto report cards. Reform educators (Zemelman, Daniels, and Hyde 1998) have argued that summative assessments, in part because they occur *after* learning was supposed to occur, have limited value.

Now, a smarter assessment type is getting wider use in classrooms. These *formative* assessments are given *during* a course of study in order to gauge students' progress and to help teachers plan instruction. Although summative assessments often acted as a punishment or as a reward, formative assessments form a roadmap to guide students toward goals. Researchers have synthesized the results of 250 sources to measure the effect of formative assessment on student learning (Black and Wiliam 1998a). They found that if teachers across the nation incorporated formative assessment, the overall effect would raise a middle-of-the-pack nation like the United States into a top-five country in the TIMSS (Black and Wiliam 1998b).

In many ways, teachers are already masters of formative assessment. We can tell within minutes of beginning an activity if children find it interesting or not. Listening to students tells us what they understand and what confuses them, and we continually adjust our teaching based on this feedback. This chapter describes some more formal formative assessments to go with these informal ones.

Also, summative assessments are being redesigned to provide feedback that is just as valuable. Reflecting on the results of after-teaching allows us to examine our own instruction and to make adjustments for the next units of study. For this reason, I will also address the use of different summative assessments. Finally, I will show how standardized tests can be an important source of both formative and summative assessments.

ASSESSMENT STRATEGIES

Three assessment strategies that are useful in the workshop classroom are interviews, embedded assessment, and artifact assessment. This assessment classification system was adapted from Foster (1999a). Each strategy can be used as either formative or summative assessment.

Interviews

One of the most valuable assessments in the workshop classroom is the interview. In this one-on-one meeting, the teacher presents a task for the student to perform. The teacher watches the child complete the task, often probing to find out the child's reasoning and asking the child to clarify or expand on his or her responses. In part because of this dialogue, interviews are highly reliable; guessing is eliminated and the teacher can get a very clear picture of each student's grasp of a given topic.

Interviews can measure fact and procedural fluency, conceptual understanding, and cognitive development.

Fact and Procedural Fluency: Addition Facts   As discussed earlier, math fact fluency with understanding usually comes through a combination of two processes. First, children commit some facts to memory as they work with these facts. Second, children become extremely efficient with their problem-solving strategies. An interview that tests for fact fluency, therefore, really tests a combination of fact recall and procedural effectiveness.

Computer Versus Paper and Pencil   This interview to test children's fluency with the addition facts is computer-based. Computer-aided assessment has advantages over paper-and-pencil tests. Years ago I assessed my students by giving them 8 minutes to finish a sheet of 100 facts. This whole-class test had several limitations. Some children would add when they were supposed to subtract or rush through and skip a line of problems, leaving holes in my data. Another drawback was that from this test, I had only a raw score and had little idea of the type of fact on which children were strongest and with what facts they were struggling. In other words, this paper-and-pencil assessment was not effectively informing my instruction.

Enter the computer. Using PowerPoint—any presentation or word-processing software that allows you to flash slides at chosen time intervals will do—I created a computer-based assessment by typing one fact equation on each page (such as 4 + 4). There is a general sequence to how children learn the addition facts, so the test starts with those earlier facts (the doubles) and then moves to combinations typically learned later. This sequence is shown in Figure 9–1. Using 3 seconds as the benchmark for fact automaticity, the program fades to a new equation every 3 seconds.

It may seem contrary to give a timed test when earlier I presented evidence that frequent timed tests seem to have a negative effect on achievement. The difference lies in the purpose and frequency of the tests. In the studies by Brownell and Chazal (1935) and Kamii (2000), teachers used timed tests frequently to drill students. This interview is given only two or three times a year, not as practice but to mark progress and inform instruction.

Conducting the Interview and Using the Data   To give this assessment, call each child, one by one, to sit in front of the computer. Tell them that they are about to take an addition facts test and that soon the screen will show an addition fact, "such as 2 + 2." Instruct the child to say the answer as soon as he or she knows it and that after 3 seconds a new problem will replace the first. Stress that the problems will go fast, so they shouldn't worry if they miss a few, just to do their best. This helps put children at ease. Then begin the program, and record on a sheet of paper which ones the child gets correct and incorrect. Figure 9–1 shows a child's correct answers checked and incorrect answers left blank. The facts get progressively harder, so it is unlikely that children will correctly solve many more if they miss five in a row. So to save time and avoid student frustration, stop the program when after five straight incorrect answers, saying, "Good job, you're all done!"

These data are useful for both planning instruction and assigning grades. Aside from the raw score, which shows the general skill of individual students, you can tell where each child is stuck. For example, the child whose work was recorded in Figure 9–1 knew the doubles and tens but had trouble deriving new facts from this knowledge. This information can be used to form instructional groups of children who could benefit from similar instruction.

139

Easy Doubles	☑ 5 + 5 ☑ 3 + 3 ☑ 2 + 2		Complex	☐ 7 + 5 ☐ 8 + 4 ☐ 4 + 9 ☐ 8 + 5 ☐ 7 + 9
+ 2	☑ 2 + 6 ☑ 11 + 2		Multiple Addend	☐ 7 + 2 + 3 ☐ 2 + 3 + 8 ☐ 4 + 7 + 3 ☐ 8 + 3 + 5 + 2
Tens	☑ 8 + 2 ☐ 7 + 3 ☑ 6 + 4 ☑ 3 + 7			
Hard Doubles	☑ 4 + 4 ☑ 6 + 6 ☑ 9 + 9 ☐ 8 + 8 ☑ 7 + 7			
Derived Doubles	☑ 5 + 3 ☑ 3 + 4 ☐ 7 + 8 ☐ 5 + 6 ☐ 7 + 5			
Derived Tens	☑ 4 + 7 ☐ 3 + 8 ☐ 6 + 3			

FIGURE 9–1   Interview response sheet showing general sequences of when children learn different addition fact types.

I give this addition test to my second-grade students three times a year: September, as a beginning benchmark, January, after we have spent several months working with number combinations, and May. A similar subtraction test is given just in January and May, because children tend to learn the subtraction facts later—often deriving them from addition facts. Most first graders are ready to be given another, somewhat easier form of the addition test in January and May.

Other Assessments for Fact and Procedural Fluency   Other interviews can be given to assess fact and procedural fluency. For example, you could have each child measure a line with a ruler, use the ruler to draw a line of a given length, and estimate the length of a line. From this interview, you can tell who knows measurement facts (which side of the ruler is centimeters and which is inches), procedures (correctly using a ruler), and conceptual knowledge (estimating length). Often the most important data that come from this assessment are the incorrect ways that children use rulers. Many children, for example, place the end of the object at the 1, which results in a measurement 1 unit too long. All these data can be used to design workshops that will build on what children understand. The same interview can be given as a final assessment, to assign grades and examine your instructional sequence.

Conceptual Understanding: Life Cycles   Interviews that assess conceptual understanding are designed for more than recall of facts and procedures. If a child understands something, he is able to apply what he knows in nonroutine situations. Giving children tasks that they have done previously opens up the possibility that they can use a procedure or recite a fact that they have memorized but not understood to complete the task. A child must draw on his understanding to solve a problem that is out-of-the-ordinary.

Plant Interview   To assess my students' understanding of the plant life cycle, I devised an interview (Figure 9–2) using a made-up plant: the juju tree. The juju tree was simply a small houseplant on which I placed a plastic flower and a "fruit" made of dried Play-Doh. I used this unique plant rather than, say, a pea plant, because children's familiarity with pea plants would allow them to use their past experiences rather than their global understanding of plants to answer the questions. Could students generalize their knowledge into a new situation?

The wording of the interview questions was also designed to assess understanding. Throughout the unit, for example, we constantly talked about where seeds are made. We challenged our students to search our school garden for seeds that might be in a root, stem, or leaf. Students became very familiar with the idea that seeds are always located in the "seed

---

PLANT LIFE CYCLE TASK INTERVIEW

This is a pretend plant. I call it the juju plant.

(Point to fruit) This is its fruit. It's called a juju fruit. If I cut it open with a knife, what do you think would be in there? Anything else? Anything else?

(If he or she said "seed"): Let's say that you are right, and there is a seed in here. If we took that seed out and planted it in the ground, what would grow?

(Point to flower). This is the flower of the juju plant. If I picked this flower (pull flower off) and every time a new flower grew I picked it off right away, would the plant get any new juju fruits or not? _____

Why or why not?

FIGURE 9–2   Plant life cycle interview.

holder," whether that is a fruit, nut, or dried flower head. We always presented the idea in that order: "Seeds are made in seed holders such as fruit" rather than "In fruits there are seeds." To assess understanding, the interview includes the *opposite* question: "If I picked this fruit and cut it open, what do you think I would find in there?" Had children's thinking about fruits and seeds become reversible?

Using the Interview Data to Plan Instruction   Results of the pretest are shown in Figure 9–3. Looking at these pretest numbers and interview notes, several holes in the students' understanding stood out. These needed to be addressed in instruction. First, it seemed as if they needed to examine a variety of fruits and to discuss their similarities to help them see the function of a plant's fruit. Second, several students held the belief that seeds from one plant could grow into a different kind of plant or that a seed could be planted and just a fruit or flower would come out of the ground. Third, the main focus of the life cycle study would have to be on the connection between flowers and fruit (and subsequently seeds); very few students knew that fruit are formed in flowers, and many children held the misconception that picking flowers off of a plant stops fruit production because flower picking somehow weakens or kills the plant.

UNDERSTANDS THAT . . .	FIRST GRADERS (N = 12)	SECOND GRADERS (N = 10)
Fruits contain seeds.	7 (58%)	7 (70%)
Seeds grow into like plants.	11 (92%)	7 (70%)
Fruit forms from flowers.	3 (25%)	2 (20%)

FIGURE 9–3   Pretest results of plant life cycle interview.

	FIRST GRADERS (N = 12)		SECOND GRADERS (N = 10)	
UNDERSTANDS THAT . . .	PRETEST	POSTTEST	PRETEST	POSTTEST
Fruits contain seeds.	7 (58%)	12 (100%)	7 (70%)	10 (100%)
Seeds grow into like plants.	11 (92%)	12 (100%)	7 (70%)	10 (100%)
Fruit forms from flowers.	3 (25%)	10 (83%)	2 (20%)	7 (70%)

FIGURE 9–4   Pretest and posttest results for plant life cycle interview.

Using the Interview Data to Reflect on Instruction   Four weeks after the end of this unit, the students were again given the interview to see how their understanding had changed. Results of this delayed posttest, along with the pretest scores for comparison, are shown in Figure 9–4.

What do these results say about this unit? First, all students came to understand that fruits contain seeds and that seeds from one type of plant grow into a plant of the same type. Second, the concept that flowers are responsible for fruit production eluded 5 of 22 students. This suggests that more experiences watching flowers turn into fruit may be necessary for all children to learn this. Next year I may supplement this unit by bringing in a flowering houseplant so children can observe and describe the flower fading as the fruit emerges.

To put this into perspective, however, compare these first- and second-grade students' knowledge with those of the U.S. fourth graders who participated in the TIMSS. Only 37% of U.S. fourth graders correctly answered, "Seeds develop from which part of the plant: flower, leaf, root, or stem?" (International Association for the Evaluation of Educational Achievement 1995). That these workshop students performed far better in an interview that is arguably considerably more rigorous than the multiple-choice question suggests that this unit was successful.

A Hybrid Interview of Conceptual Understanding: Place Value   The beauty of an interview is that you can learn a lot in a one-on-one conversation. Teachers, however, have to balance giving the best assessment with all the other time pressures in their classroom. A hybrid interview that uses interview techniques but is given to more than one student simultaneously is better than giving a poor assessment.

1. *Assessments of place-value understanding:* Years ago, I assessed place value by giving each child a sheet of paper with the number 58 written on it. Then I instructed them, "Underline the number in the tens place. Circle the number in the ones place." Although I then knew who could *identify* the place-value places, this didn't tell who really *understood* place value.

   An excellent interview that tests for this understanding was developed by Kamii (1989). I have used this with my second graders to see who has a sound understanding of 10, and have found that it is highly reliable. In other words, students who can accurately complete this task show other signs of understanding how larger numbers are organized.

   The first year I gave this interview, I was surprised at my second graders end-of-the-year performance. Only five of twenty-four students (21%) passed this test. This compared very unfavorably to Kamii's (1989) data, in which 67% of the second-grade students passed. This told me that I needed to look at my instruction.

   The next year I planned a series of workshops focusing on place value. One workshop sequence, for example, had students involved in an imaginary factory in which they packed and unpacked imaginary candy (McClain, Cobb, and Bowers 1998). They spent a lot of time putting the candy pieces (Unifix cubes) into rolls of ten, counting them, repacking them in other ways, and solving problems involving sending out orders of candy. All this activity was followed by reflection—discussing and writing about our base-10 system of candy packing, especially the relationships between quantity, number symbols, and number words (see Fuson, Smith, and Cicero (1997) for a discussion). That this wrestling with place value led students to understanding was shown in the year-end task interview data; 21 of 23 (92%) passed the Kamii test.

2. *Teacher in a time crunch: Developing a hybrid interview:* One disadvantage of one-on-one task interviews is that they can be time-consuming. The Kamii test, for example, can take 5 minutes for each child, which in a class of 25 represents a good portion of the school day in which I have to simultaneously conduct interviews and monitor the rest of the class. One May, I found myself in a severe time crunch. I had to finish our last science unit, students needed assistance editing their mystery stories, and report cards were right around the corner. I decided that I was unable to conduct the one-

on-one interview but still wanted an accurate reading of their place value understanding. So, I instead converted the interview from the original Kamii test to a whole-class format. This is shown in Figure 9–5.

3. *Conducting a hybrid interview:* This assessment is conducted by presenting an overhead of Figure 9–5, with the 2 written in red marker and the 3 in blue. Each of the children is given a paper version, with the numbers colored similarly. Then begin, reading this script:

   1. On this sheet there are 23 circles, and that number is printed below the circle.

   2. With a blue marker, color in the number of chips that *this* part of 23 means (point to and underline the blue 3.)

   3. Now, with your red marker, color in the number of chips that *this* part of 23 means. (Here point to and underline the red 2.)

   4. Now, look at the chips you colored red. On the lines at the bottom, I want you to write *why* you colored that many red. What is it about this part of 23 (pointing to the red 2) that made you color that many red?

   Typically, almost no beginning-of-the-year second-grade students can pass this assessment. The response in Figure 9–6(a) is typical of a student who sees the 2 in 23 as 2 ones. As students become more secure in their class-inclusion development, they are better able to understand that the 2 in 23 has a double meaning: 2 tens and 20 (Figure 9–6(b)).

4. *Limitations and benefits:* Clearly there are limitations to these hybrid interviews. First, the teacher can't ask for clarification or expansion of a student's response. Second, because probing student thinking is as important as their answers, students need to be able to write well enough to express their ideas. This may not happen in many students before second grade, and some students may not be able to do it at all in the primary grades.

   That said, a well-designed hybrid interview is a workable alternative when time is limited. I have used both this whole-class assessment and the original Kamii interview for several years and have found an acceptable level of correspondence between the two versions. When there is reason to suspect that a child does not understand what was being asked during the whole-class task, I follow up with the one-on-one interview. For example, one year five children failed to pass the test; that is, their drawing and/or their writing were inadequate. Of these five, two colored only one circle red,

MATH WORKSHOP

Name _____

Date _____

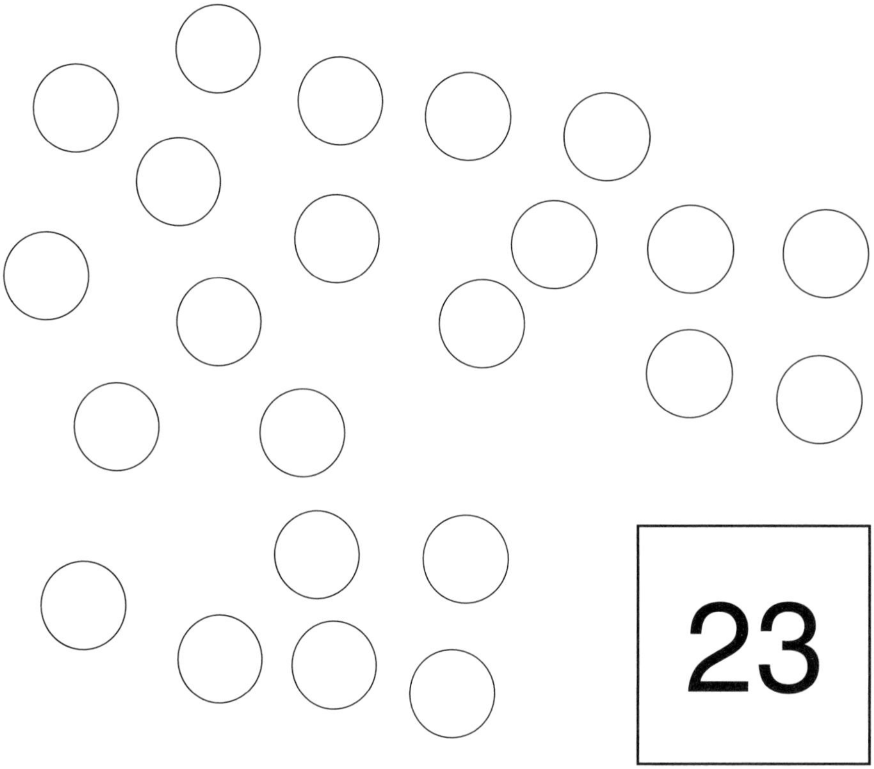

23

_____

_____

_____

FIGURE 9–5 Hybrid interview student-response sheet for place value.

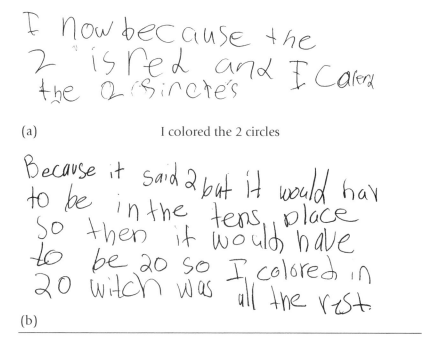

(a)   I colored the 2 circles

(b)

FIGURE 9–6   Unsuccessful and successful response sheet for place value.

two (who were developing writers) wrote unclear or inadequate explanations, and another child (who was an advanced writer) failed to write anything. To these students I gave one-on-one interviews. Although not as telling as the individual interview, this hybrid format is a quick way to assess understanding of place value.

Assessing Cognitive Development: Student-Directed Workshop   A child's cognitive development is his or her collection of mental tools that he or she uses to make sense of the world. Knowing what tools—called cognitive structures—children have affects what and how teachers can teach. Interviews are often the best way to assess a child's cognitive development. The definitive source for such interviews is *Structures of Thinking: Concrete Operations,* (Phillips and Phillips 1996). However, another book, called *Developing Logical Thinking in Children* (Phillips 1991) is often more useful because it presents the interviews in a less formal format that is better suited for most classroom situations.

Interviewing as Teaching in the Student-Directed Workshops   Student-directed workshops are a perfect venue for conducting these informal interviews of cognitive development. Students are working by themselves, with objects that interest them, usually doing something that connects directly with one of three of the primary developmental structures: classification, class inclusion, and conservation of number. Children are often involved in these behaviors anyway, so asking questions about what they are doing is a natural. Because these interviews are given during classroom activities, this interview type is also considered an embedded assessment, which is the topic of the next section.

It is also important to note that the questions that comprise the interviews are *the same questions* that are likely to help children attain cognitive structures. If a child is developmentally ready, the interview questions could encourage the reflection needed to further develop these understandings. In this way, these interviews assess and teach at the same time.

1. *Conducting the interviews:* These interviews can be conducted with most of the objects that children choose to work on during student-directed workshops. I save these interviews mostly for kindergartners, first graders, and those second graders who seem to lack a good understanding of number. Research (Phillips and Phillips 1996; Hiebert, Carpenter, and Moser 1982; Heuser and Foster, in preparation) shows that these groups are often still developing the primary cognitive structures.

2. *Classification:* If a child is grouping objects in any way, it is easy to conduct an informal classification interview. Children readily talk and rearrange objects in response to questions such as, How did you group your objects? Can you do it another way? "Can you group them so that you have three groups instead of four (or two instead of four)?" What they say and do shows their ability to classify objects using set criteria (type, color, size, etc.), to reclassify objects using new criteria, and to combine or separate like groups—all important processes that help children understand math and science.

3. *Class inclusion:* Class inclusion is developed after classification. It deals with the relationships between groups and subgroups. The class inclusion interviews given in Phillips (1991) and Phillips and Phillips (1996) can also be given as a child groups objects.

4. *Conservation of number:* To assess number conservation, look for situations where a child has set up rows of objects that correspond one-to-one. This most often happens as children arrange objects in graphic designs or set up objects (such as dinosaurs or airplanes) for a "fight." Again, the interviews in Phillips (1991) and Phillips and Phillips (1996) fit in naturally as students engage in these common behaviors.

Embedded Assessment

A second type of assessment is embedded assessment (Foster and Heiting 1994). In embedded assessment, assessment and instruction are not separate processes. Instead, children are assessed while they are involved in learning activities. An obvious benefit to embedding assessment into instruction is that less time is needed for giving tests, so more time can be devoted to teaching.

Assessing Strategies During a Math Game   Watching pairs of students as they play Double Roll and Add is a good way to assess children's progress during the addition fact unit. Clipboard in hand, I sit next to the children for several turns, looking for the different strategies that they use to determine who won each roll. I get more information from the kids by asking them how they know they won (or lost), although with some children their strategies are apparent, such as those who physically touch the die dots as they count up.

As I find out what strategies each child uses and how effective they are with those strategies, I can often respond immediately. Matthew, for example, is a second grader who often still counts all; that is, after rolling a 6 and a 5, he counts every dot on both dice to figure the total. To help lead him to using the more efficient counting on strategy, I asked how many were on "this die" (the 6). "Six," he immediately responded, indicating that he perceptually knew what 6 looked like on a die. I then covered that die and said, "You know that there are 6 on this die, so how can you figure out how many you have all together without counting each and every dot?" After some thought, he counted up on the die that was showing, "7, 8, 9, 10, 11." I had him roll again to see if that strategy would work on his next roll to give him more practice.

Knowing what strategies children use helps inform the reflection period. I ask some children to present their strategies by announcing, "I noticed that Vicky had an interesting way to solve 11 + 9. Can you tell us how you did it, Vicky?" Other times I will share some strategies from which I feel some students can benefit. For example, "I saw one student count all the dots to find out his total." (Here I demonstrate with drawings of the 6 and 5 dice.) "Later I saw him just say the larger number and count up, using the dots on the die with the smaller number, like this. Which strategy do you think is better?" Questions such as "Is this strategy easier than this other one?" or "What do you think of how that child did it?" often lead to productive debate—debate that might not happen without this embedded assessment.

Listening During Plant Inquiry   "The best way to support inquiry is to obtain information about students while they are actually engaged in science investigations with a view toward helping them develop their understandings of both subject matter and procedure" (National Research Council 2001, 3). As the students presented their plant inquiry plans, they were performing an essential part of the inquiry process. Additionally, assessment is embedded in these presentations; discussion generated by the audience often shows very clearly what students understand and what still eludes them. I told the students that they should try to think of one question and one potential problem with each inquiry in order to encourage this discussion.

Students shared their questions and comments, and I recorded those that provided insight into students' thinking on a sheet of paper. I stopped and helped students process comments that I thought would help them understand inquiry better. For example, on the poster entitled "Can a plant survive in the dark?" the presenters noted that they would put their plant in a dark closet but that they would water it every day. "But when you water it," one child explained, "it will get some light, so it's not really in the dark all of the time." "How could they avoid this problem?" I asked, and another student suggested that they turn off all the lights before they open the closet. Similar problems are encountered all the time by investigators, and students need to learn how to overcome them.

Another child asked what would happen if a group accidentally dropped their plant and it broke, thus ruining the experiment. "We could plant two plants, in case one breaks," was the reply, and I used this comment to tell the children how experiments usually have duplicate setups to avoid this problem.

One boy's comments during these presentations surprised me. Khalkedan has trouble completing tasks. I could tell little about his knowledge of inquiry from his poster, because he was wandering around the room while his partner was left to do most of the work. Although written work was not his strong point, what he knew about inquiry stood out clearly during these presentations. In one instance, presenters proposed planting bean seed in a small pot and a pea seed in a large pot to determine what size pot would best grow plants. "Why would you plant different kinds of plants?" he asked. "To be fair, they should both be the same kind." More than any other child, Khalkedan kept coming back to this idea of a fair test, a concept critical to inquiry that eventually leads to the control of variables (National Research Council 1996). I would have remained unaware of his inquiry ability without my notes taken during this embedded assessment.

Student-Directed Workshops   Embedded assessment is built into every student-directed workshop. Observing children and probing into their thinking as they work is an excellent way to give formative feedback. I find it cumbersome to record my observations of children as they work, so instead I can focus on providing immediate feedback.

Artifact Assessment

Artifacts from activities—including observation journals, problem-solving sheets, and inquiry plans—can be assessed. Artifact assessment is a type of performance assessment because almost all workshop activities require children to apply (as opposed to simply recall) knowledge.

Math Problem-Solving Sheets   As noted in Chapter 4, problem-solving sheets are often done collaboratively by students. Once every other week or so, however, I have students complete them individually. These sheets act as tests of students' progress. During these times the children know to abide by certain test-taking rules: to set up their own "offices" of two folders for privacy, not to talk or look at other papers, and to quietly draw on the back of their papers when they get done. With younger children, these rules create a problem if they cannot read the question, so I usually read the questions aloud, pausing for enough time between questions to let students solve each one.

Evaluating these sheets usually reveals two things about each student: what strategies he or she is using and how many answers are reasonable and correct.

What Strategies Are Being Used?   Figure 9–7 shows some responses to the question, "If 3 pigs live in every house and there are 6 houses in the village, how many pigs live in the village?" Each response shows a range of strategies for solving the problem, from simple to complex; (a) shows a very literal direct modeling of the story. The child who did (b) also directly modeled the problem but created an abstract model of the village, and (c) shows the more efficient counting strategy. In (d) the child used number facts but didn't understand the context of the problem and thus got an incorrect answer. The child in (e) correctly used a series of number facts to solve the problem.

Normally I do not record each child's strategy but instead bring up certain strategies in the minilesson of the next workshop to illustrate points. For instance, in the next minilesson in this sequence I showed (d)—without attaching a name to it, of course—and asked what mistake the student made. One student replied that the answer didn't make sense and that the student who made this mistake needed to think about what the story was really about. This led to further discussion around a frequent theme in my classroom—that math needs to make sense. I also presented (a) and (b) and asked which one was easier to do. Most chil-

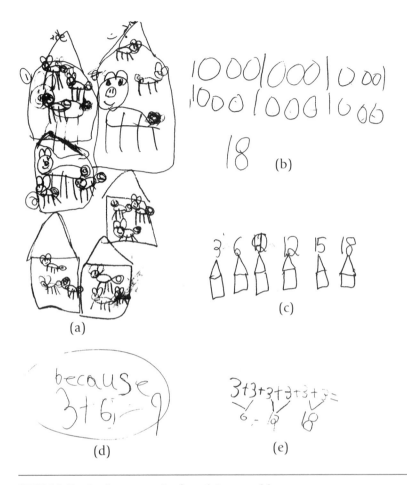

FIGURE 9–7   Student strategies for solving a problem.

dren concluded that (b) was a much easier strategy. In this way I hoped to encourage those students who regularly drew detailed drawings to, when they were ready, adopt less detailed representations.

What Answers Are Reasonable and Correct?   Some answers are wrong but reasonable. I record both the number of reasonable but incorrect answers and the number of correct answers for each child. Looking at these scores shows me which children are really struggling with the concepts versus which children are just making minor errors. I can then work with each group on their own issues. These scores indicate whether to move on to a new topic or to work more on the old. They also come in handy when reporting to parents.

	4 EXCELLENT	3 GOOD	2 OK	1 YOU DIDN'T EVEN TRY!
Accurate skin and body details?	Lots of detail	Some detail	A little detail	No detail
Big drawing?	Big	Medium	Small	Tiny
Focus on tadpole, not its surroundings?	No surroundings	A little of the surroundings	Some surroundings	All surroundings
Body parts labeled?	All parts labeled	Most of the parts labeled	A few of the parts labeled	No parts labeled
Words that tell size?	Lots of detail	Some detail	A little detail	Didn't talk about size
Words that tell how it changed?	Lots of detail	Some detail	A little detail	Didn't talk about how it changed
Number in each column	___ × 4	___ × 3	___ × 2	___ × 1
Totals	_____	_____	_____	_____
Grand Total	_____			
KEY	22–24 Excellent	18–21 Good	12–17 OK	6–11 You can do better!

FIGURE 9–8   Rubrics for assessing tadpole observation logs.

Students Self-Assessing Tadpole Journals   Having students assess their own work is a powerful way to both assess and teach (Marzano 2000). When children ask critical questions—Am I doing my best? How does this work compare to excellent work? Is there a way that I can make this better? Do I care about what I am doing?—they can develop insights into their own thinking and learn to judge their understanding of math and science. Student self-assessment is especially powerful when students help to develop the standards to which their work will be judged. This section describes how students can learn to self-assess and at the same time develop their abilities to observe and describe.

Analyzing Tadpole Drawings   As my students monitored the progress of their tadpoles' development, they made drawings and notes in their observation logs. Students' first attempts in this log left a lot of room for improvement. Drawings of the gray tadpoles were done in red marker outline, so small that showing any detail was impossible. Students made elaborate drawings of the cup in which the tadpole swam or sometimes transformed their tadpoles into imaginary settings. Tadpoles always were drawn smiling; one would conclude that tadpoles are among the cheeriest creatures in the animal kingdom based on these initial observation-log entries.

Systematic observations and accurate descriptions are essential to the inquiry process. Looking at these early drawings, however, I concluded that students did not know what a quality observation-log entry looked like. I began a process of communicating the standards for quality work by showing students a drawing from each child. I told the students that for each drawing, they were to come up with one strong point and one area for improvement. I did the first few for them, for example: "I like how Brian drew the shape of his tadpole. It looks just like the tadpole, especially around the tail. One way that he can make this better is to draw the mouth so that it looks just like that tadpole mouth. Did anyone see their tadpole smile? If you look real closely, you can see that their mouths are small black circles. I know you want your tadpoles to be happy, but scientists try to make these observational drawings look exactly like what they see."

Soon the students took over. I was surprised at both how generous they were with their praise, and how accurate their criticisms were. After this half-hour reflection students had a much better idea of how a good observational drawing should look.

Writing Rubrics   Now, it was time to put this knowledge into writing. I set up the bones of the rubric system for them (based on Marzano (2000)), summarizing our discussion on what makes a good observational drawings as I filled in the left-hand column of the grid shown in Figure 9–8. I also set up the point system on the top row, but once we began the process, students became quite adept at providing descriptors for each rubric.

Using the Rubrics   In the next day's workshop, I gave each child a copy of the self-assessment. We reviewed the rubrics, and then children began observing the tadpoles and completing their observational logs. I conferenced with children as they drew and wrote: "Is your skin color accurate? What would you give yourself right now on your labeling? I think that you can get an excellent on this; what's the one area where you can do it better?"

After their logs were complete, children reflected by doing the self-assessment. Students worked on the math in pairs—most were proud to be multiplying—and then students turned them in to let me have a shot at grading them. I wondered how honest or accurate students would be grading their own work. I found that in almost every case, the children's scores either matched mine or were slightly lower, indicating that they could accurately judge their performance. The value of this self-assessment also showed in their logs. Compare, for example, the "before" self-assessment and "after" self-assessment entries shown in Figure 9–9.

Plant Inquiry Posters   In addition to the notes I took during the presentations of the plant inquiry posters, I wanted to assess students' ability to plan and draw a fair-test experiment in poster form. I wrote a set of rubrics using the general rubric format suggested by Marzano (2000) in order to judge the posters as objectively as possible. These rubrics are shown in Figure 9–10, along with the number of posters made by pairs of students that fit into each of the five levels.

These data suggested that most children were able to plan a fair-test experiment but that half of these children's posters included important details, whereas the other half did not show details that would have been needed to complete the experiment successfully. One example of a detailed poster is shown in Figure 9–11(a). Notice the measuring cup indicating how much water the plants should be given. Another pair of students noted, "We will look at the color (of the plant) to see if it is dead." An example of a poster lacking detail is shown in (d). Although the drawing shows the experimental setup, which could answer the question, "Is a big pot or a little pot better?," it didn't communicate the procedure after the initial setup. Would the plants have to be measured? Would you have to wait for one plant to die?

Obviously, students would benefit from more support in thinking through and then drawing and writing these details. I shared my thoughts with my teaching partner, and together we set up a simple sheet that asked students to sketch out their setup *and* write what they would then have to do during the course of the experiment. Nearly all of her students' posters included enough detail to be judged Level 4 on the rubric in part because of this instructional modification.

Adapting Standardized Tests

No chapter on assessment would be complete without a word on standardized tests. "Employing standardized achievement tests to ascertain educational quality," writes Popham (1999, 15), "is like measuring temperature with a tablespoon." There are many good arguments against the way standardized tests are used in our educational system. Among those

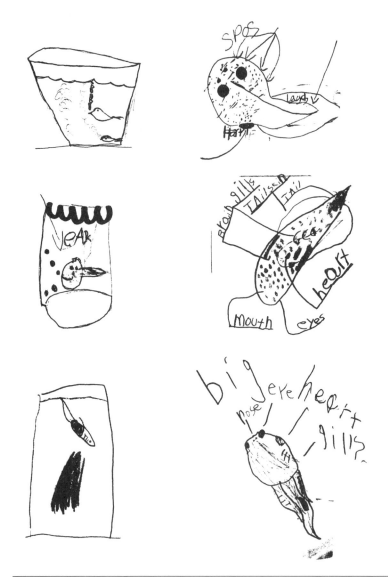

FIGURE 9–9   Tadpole drawings before (left) and after (right) student self-analysis.

is that, specifically in regard to math, standardized tests merely assess lower-order skills, such as the ability to recall simple facts, apply standard algorithms, and perform routine procedures (Romberg 2000).

Although this argument does have a grain of truth, at least two standardized tests—the NAEP and TIMSS—have blocks of questions that are complex and nonroutine. These are specifically designed to address the mathematical and scientific understanding. Quite a few of them have

LEVEL	STUDENT PAIRS . . .	NUMBER OF STUDENT PAIRS
4	Can plan a simple fair-test experiment and present it in detail.	5
3	Can plan a simple fair-test experiment, but presentation lacks many details.	5
2	Can plan a simple experiment, but it has "unfair" as well as "fair" aspects.	2
1	Cannot plan a simple fair-test experiment.	0
0	Poster provided little or no information in which to make a judgment.	0

FIGURE 9–10 Rubrics for assessing tadpole observation logs.

been released to the public and are readily available online.[1] Using these questions has powerful potential in the workshop classroom. They can be a source for problems for students to solve during workshops, or reflection can center on different ways to solve the different problems.

Given in the form of interviews or on problem-solving sheets, these problems can also be used as assessments. Borrowing questions from standardized-test questions to assess your children has two big benefits. First, their results are normed and easily accessed.[2] Comparison to national or local results can give further feedback on your students' learning and your instructional practices.

Another benefit of using these questions as assessments is that parents tend to assign a higher worth to standardized-test items than to teacher-created assessments. For this reason, I occasionally publish different test questions in my weekly newsletter, comparing the results of my class with the U.S. fourth-grade population. See Figure 10–3 for an example. Even though my first and second graders are 2 to 3 years younger than the fourth graders who took these tests, they often meet or beat the performance of the older students. When a smaller difference in age between my students and the control group is useful, I often use the TIMSS data for Singapore. I explain in the newsletter that this nation outper-

1. Released test items from the NAEP are available at <http://nces.ed.gov/nationsreportcard/itmrls/pickone.asp>. TIMSS-released questions are available at <http://www.timss.org/TIMSS1/items.html>.

2. Results from NAEP tests are available at <http://nces.ed.gov/nationsreportcard/itmrls/pickone.asp>. TIMSS results are available at <http://isc.bc.edu/timss1995i/data_almanacs_95.html>.

Is old Coffee better then new Coffee?

(a)

How high will a plant grow if you give it Koolade?....

(b)

FIGURE 9–11 Inquiry posters.

What will happen if we burn a seed and plant it? Will it grow?

(c)

Is a big or littel Pot better?

(d)

FIGURE 9–11   Continued.

Four children measured the width of a room by counting how many paces it took to cross it. The chart shows their measurements.

Name	Number of Paces
Stephen	10
Erlane	8
Ana	9
Carlos	7

Who had the longest pace?

A. Stephen

B. Erlane

C. Ana

D. Carlos

Reproduced from the TIMSS Population Item Pool.
Copyright 1994 by IEA, The Hague

FIGURE 9–12   Questions from TIMSS.

formed the United States on the TIMSS and that their students took the test in third grade instead of fourth.

Example

To judge my second graders' understanding of linear measurement, I give them two questions, one from the TIMSS (Figure 9–12) and one from the NAEP (Figure 9–13). The questions were given as part of a problem-solving sheet at the end of our study of linear measurement. Results from last year are shown in Figure 9–14.

These assessments give a summative picture of linear measurement understanding. I used each student's performance on these tasks to help assign grades on the report card. Additionally, looking at the whole-class data in Figure 9–14 supported the effectiveness of the instructional sequence that I used to teach measurement. It is obvious that the workshop students developed a far greater understanding of linear measurement

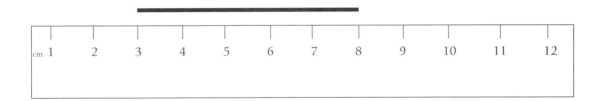

How long is this line segment?

❑ 3 cm

❑ 5 cm

❑ 6 cm

❑ 8 cm

❑ 11 cm

❑ I don't know

FIGURE 9–13   Questions from NAEP. Reproduced from Lindquist and Kouba 1989. Copyright by NCTM.

	QUESTION 1 (TIMSS)[1]	QUESTION 2 (NAEP)[2]
Workshop Students	58% (Grade 2)	85% (Grade 2)
U.S. students	10% (Grade 4)	14% (Grade 3)
		49% (Grade 7)

[1] International Association for the Evaluation of Educational Achievement 1995, 1998.

[2] Lindquist and Kouba 1989

FIGURE 9–14   Comparison between workshop second-grade students and U.S. third-, fourth-, and seventh-grade students.

than did the students taught in other ways, even though the nonworkshop students were in grades 3, 4, and 7. Note that even seventh graders, many of whom likely received several years of traditional measurement instruction, were outperformed by the workshop students 5 years their junior. These data offer further evidence of the advantages inherent in the workshop format.

10

Communicating with Parents and Administrators

Earlier, I hypothesized that two main factors may be working against teachers who try to adopt reform math and science techniques. The first is overcoming the difficult challenge of making significant, sustained changes in teaching behavior. The first nine chapters of this book were about making these changes by teaching math and science through the familiar workshop format. This chapter addresses the second factor: convincing administrators and parents about the benefits of teaching that is unconventional. Following are five strategies to get parents and administrators on your side as you work to teach math and science workshops.

CURRICULUM NIGHT: TEACHING IS DIFFERENT . . . AND THAT'S A GOOD THING!

Every September my school has a curriculum night, when parents come to meet their children's teachers and to hear about the upcoming year. Because learning math and science in a workshop classroom looks quite different from learning them in a traditional one, this night is the best time to state very clearly that (1) it *is* going to be different, and (2) it is a *very good thing* that it is going to be different.

In my 20-minute speech to the parents, I touch on many of the ideas in this book:

• The world is different than when you grew up; every child now needs to have a deep understanding of math and science.

• Many adults in the United States don't like math or science and freely admit that they are not good in these subjects. Parrot teaching methods are in part to blame for this. The poor condition of math and science education was so alarming that business leaders, teachers, educational researchers, scientists, and mathematicians got together to produce national standards.

- Workshop math and science is based on these standards and on research that shows what really works in the classroom. Traditional methods often are not. We demand that our doctors be up-to-date with best practice in medicine. Why should we demand anything less from our teachers?

- In workshop math and science your child will be doing things that look different than traditional math and science: working a lot with manipulatives, choosing among different activities and solution methods, talking and writing about math and science, and tackling real-world problems and interesting questions.

- Your child will enjoy workshop math and science (although it may take him or her some time to get used to it) and will thrive academically.

I also use this time to explain what parents can do to help. First and foremost, I ask parents *not* to show their children how to solve problems. As you can imagine, this is a real attention-grabber. To explain, I display a table of data showing second-grade children's responses as they solved the problem 7 + 52 + 186 mentally, without paper or pencil (Kamii and Dominick 1998). This table compares three groups of students. I first share the student answers from two of the groups: (1) children who were taught the standard addition algorithm by their teacher and (2) children that were *not* taught this algorithm by their teacher and whose teacher called each of the children's parents to ask them not to teach the algorithm either. The data show that nearly four times as many children in the no-algorithm group as in the algorithm group figured the correct answer (245). Additionally, the incorrect answers in the no-algorithm group were, in most cases, very reasonable—between 235 and 255. In the algorithm group, most incorrect answers were so unreasonable (between 9308 and 29) as to suggest that these children did not understand what these numbers meant.

I then display the third column. These children's performance falls in the middle of these first two groups, both in the number of correct answers and in the reasonableness of their errors. "These children, " I explain, "had a teacher that did not teach them the algorithms, *but he forgot to ask his students' parents not to teach them the algorithm.* So that is what I'm asking you now: please do not show your child how to solve problems."

This up-front approach to informing parents has been very successful for me. Being proactive has helped me avoid the urgent phone calls demanding, "How in the world are you teaching my child?" The direct approach also seems to leave an impression; often months after curriculum night, parents will say, "I remember how you asked us not to show her how to solve problems." To help reinforce and expand on curriculum night, I send home the handout shown in Figure 10–1.

163

Helping Your Child Succeed in Math

Family Math Night

Do play math games with your child.

Along with reading aloud to your child, playing math games is one of the best things you can do for your child's education. Good math games teach important concepts and skills, sometimes without your child even knowing that they are about math. Game playing is a fun, social experience likely to give children a positive attitude about math. And research shows that children learn math facts and computation better through games than through traditional worksheets.

Do involve your child in everyday uses of math.

When you involve your child in everyday uses of math, you show that math both makes sense and is useful. Real projects and interesting problems help your child grow important math ideas. Ask your child to equally share a pile of candy with his friend. That's division. Have him help you with a simple building project, and measurement and geometry come to

life. When she tries to guess how much the grocery bill will be, she is estimating. Children sort as they organize their trading cards or toys, multiply when they figure out how many feet are in their family, and count as they find out how many steps it is from their bedroom to the kitchen.

Do talk math with your child.

Your child's feelings about math are shaped in part by what you say about it. Parents who say "I'm not a math person" may give their child the idea that some people can "get" math while others can't. On the other hand, if you say "Math was always easy for me," a child struggling in math may feel that there is something wrong with him. The best message you can send is that sometimes math will be easy, but like many other worthwhile things in life, math can sometimes also be very hard. But, if you work hard enough and try to understand, the odds are that you will succeed at even the most difficult parts.

FIGURE 10–1   Curriculum night handout.

MATH NIGHT: TEACHING ALTERNATIVES TO SKILL AND DRILL

Parents need frequent gentle reminders about what *not* to do to help their child in math. But many also appreciate a hands-on lesson on what they *can* do. A family math night is a great way to do this. Last year teachers at my school planned and staffed our own family math night. It was organized around six different math strands: numeration, basic facts/operations, patterns, geometry, money, and time. For each strand there was a room set up with all the materials needed to do different activities that addressed that strand. Most often the activities were games. Children brought their parents from room to room and taught them how to do the activities.

The night was a big success. Children were excited to show their parents what they do every day in math. Parents gained both a sense of what our math program was all about and some concrete ideas of how to help

Don't show your child how to solve math problems.

Children have special ways to solve problems that don't always make sense to adults. At the same time, children don't usually understand adult problem-solving methods. Therefore, it is important to resist the temptation to show your child how you would solve a problem. Instead, encourage him to use his own problem-solving strategies. What should you do then, when your child asks "How can I add 17 and 9?" Often the best response is "How do *you* think you can do it?" It also may help to put problems into a real-life context: "Well, if you had 17 pennies and I gave you 9 more, how many would that be?" Asking if he has ever solved a similar problem might help him use a past strategy. Finally, you can ask, "If you got some pennies or drew a picture, would that help you?"

Don't encourage your child to memorize the math facts (3+4=7, 2x6=12, etc.)

While it is very important that children develop automatic knowledge of math facts, the best way to do this is *not* by memorizing them. Like memorized problem-solving methods, math facts that are memorized are often forgotten. Additionally, time spent memorizing is time in which children are not thinking about numbers, place value, and operations $(+,-,x,\div)$. Children who are encouraged to wrestle with a problem like 13 - 6 develop strategies like counting down ("12,11,10,9,8,7—I counted down 6 numbers"), subtracting the number in easier parts ("13-3=10, and 10-3=7") and using known addition facts ("6+7=13, so 13-6=7").

Games are one of the best ways to learn the math facts at home. These will be coming home at suitable times from your child's teacher. It is also very helpful to ask children to explain their thinking. For example, pose a problem at an appropriate level for your child ("What's 2+4?"), then ask,

"How did you figure that out?" Children who regularly explain their thinking strategies and play math games learn the math facts far better than do children who just memorize.

Don't use worksheets, flashcards, or timed tests.

You know that raising a child is a lot of work. Aside from their physical, emotional, and social needs, parents need to actively participate in their child's education to ensure success. This pamphlet gives several suggestions on how you can best do this in the area of math. Between talking about math, playing math games, posing interesting math problems, and doing real-life projects involving math, there will be little time left in the day for other activities. While your child's teacher may occasionally use worksheets, flashcards, and timed tests, we recommend that you do not. Enjoy your time with your child!

© Daniel Heuser, 2001 (773 878 3445)

FIGURE 10–1   Continued

their children succeed. To further encourage parents to do these worthwhile activities in place of flashcards and workbooks, each family was sent home with a math night gift. Businesses in the community and our local PTA donated playing cards, dice, and rulers, so that every family would have some of the materials needed to do the math night activities at home. Later in the year I sent home a reminder of some of the games (Figure 10–2). One resource that you may find helpful for organizing your own family math night is available online (Everyday Learning Corporation 2000).

NEWS FLASH! ADVERTISE YOUR SUCCESSES

Communicating the effectiveness of your math and science programs to administrators and parents is one key to a successful workshop classroom. In my weekly newsletter I frequently send home results of assess-

MATH GAMES FOR LEARNING ADDITION AND SUBTRACTION FACTS

Playing math games is one of the best ways for children to learn the basic addition and subtraction facts (5+4, 13-8, etc.). Games are fun, and children learn the math facts better by playing games than through worksheets or flashcards.

DOUBLE ROLL AND GRAB

- *Materials*: 4 standard dice, a big pile of pennies.
- *Rules*: The players roll their two dice at the same time. The player with the higher total takes that number of pennies from the "bank" and puts it in his winner pile. The other player takes no counters. If both players roll the same total they both take that number of chips. Play continues until all of the counters in the bank are gone. The player with the most number in her winner pile wins the game.
- *Variations*: You can purchase or make dice with larger numbers, or use three or four standard dice to make the game more challenging.

We continue to urge you **not to have your child memorize the facts without thinking**, but instead to help your child develop *strategies* for adding and subtracting. Playing these games is a good way to do this. It also helps when you ask him how he figures out the problems. Prompts such as "How do you know that 8 + 9 = 17?" or "I think that my 7–4 beats your 7–5. Do you think I'm right?" help children learn strategies. HAVE FUN!

©2002 by Daniel Heuser from *Reworking the Workshop*. Portsmouth, NH: Heinemann.

FIGURE 10-2   Addition game handout.

ments and student work samples to show how much my students are learning. And I always send a copy to my principal. One example is shown in Figure 10–3.

GET INVOLVED: SHAPING CURRICULUM AND DELIVERY

Teaching workshop math and science is much easier when your school has a sound curriculum. On the other hand, you will not get much encouragement to use workshops if the goals handed to you are developmentally inappropriate or call for rote memorization and low-level skills development.

This question is from the Third International Mathematics and Science Study, an international standardized test given across the world to nine- and thirteen-year-old children.

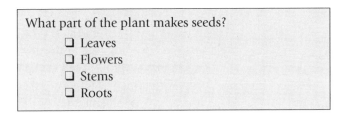

What part of the plant makes seeds?
- ❑ Leaves
- ❑ Flowers
- ❑ Stems
- ❑ Roots

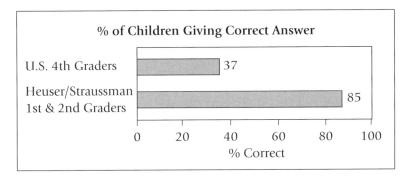

This is a graph showing how our children's performance compares to the national average.

How did your first- and second-grade children do better on this question than students across the nation who were two and three years older? Part of this success probably stems from how your child is taught science:

- ❑ Your child learned *hands-on*, examining many, many plants in our science garden and at home. This lets them use their senses to engage their minds. Hands-on learning is not the norm in classrooms across the country.

- ❑ Your child was able to choose between different problems to solve and questions to explore. *Choice* motivates children and lets them learn in ways that best suits them. In many U.S. classrooms, all children are expected to do the same activities.

- ❑ Your child wrote and talked about plants every day. This communication helps children *reflect* on their experiences and build new understanding. Science time in many other U.S. classrooms is dominated by teacher-talk, with little opportunity for children to communicate their own ideas.

- ❑ Your child learned through *inquiry*, as he or she designed their own experiments, interpreted the results, and presented their findings to the class. This helps children think like scientists, using both what they observe and what others know to understand science. Many teachers use experiments, but usually it is an experiment designed by the teacher, with the teacher telling students what the results mean.

Children learn a lot more when education is matched to research on how they learn best. Even though our children did very well on this question, it doesn't really show just how much they learned. Most children passed a far more rigorous test, in which they had to solve a complex problem and justify their answers.

FIGURE 10–3   Classroom newsletter.

I am very fortunate to work in a district that periodically reflects on and revises our curriculum. I am also very fortunate that teachers are given a leading role in this process, working side by side with principals and other administrators. Last year I served on a math committee that realigned our curriculum with the national standards. This role gave me the chance to shape the curriculum to match research and best practice.

One of our tasks was to adjust grade-level expectations to reflect what children could really do. For example, the original first grade expectation for addition was confined to problems that could be solved without regrouping (e.g., 22 + 6). This seemed to play into the notion that this type of problem should be solved by the standard algorithm, and since first graders have difficulty memorizing how to "carry the ones," problems requiring regrouping are beyond their reach. It was clear, however, that even beginning-of-the-year first graders can solve addition with and without regrouping if they are encouraged to use strategies that make sense to them. So the benchmark was rewritten, "Add and subtract two- and one-digit numbers (e.g., 17 + 4, 15 − 8) using appropriate strategies other than borrowing and carrying."

My district is also presently examining our science program, making adjustments to match best practice in science education. My committee's job was to summarize just what this best practice was. We wrote and presented the document in Figure 10–4. This is being used as a guide in writing curriculum, choosing materials, and promoting teaching techniques.

Getting involved in shaping your local math and science programs can be a lot of work, and one person's participation is no guarantee that the final results will be sound. Ultimately, however, your job will be much easier if you are teaching in an environment where reform math and science is valued.

BOOK CLUBS

Almost everywhere, it seems, people are getting together to talk about books. Book clubs are a social way to learn, and teachers learn together as they read, discuss, and try out ideas from good books about teaching. Include your principal in the club, and the process of improving math and science instruction becomes much easier. Studying together is a powerful binding force. Principals bring expertise that is helpful in facilitating discussions and change, and inviting your principal into your room to observe workshops and provide feedback cements that partnership.

Book club participants set a schedule in which to read the book and to pick certain strategies or activities that they would like to try in their classrooms. The club then meets periodically to share what they tried. Discussion revolves around what went well, what problems arose, and

BEST PRACTICE IN SCIENCE EDUCATION

In this new century of technology and information, there is an unprecedented need for a scientifically literate society. The changes in science education needed to meet this goal are considerable. Fortunately, we do not have to guess at how to best prepare our children—there are already educational methods that are supported by substantial amounts of research. These proven measures are called Best Practice. Our science education program must have its foundation set in Best Practice; anything else is denying our students the tools to solve the pressing personal and societal issues awaiting them in their lifetimes.

1. **Understanding**. All students need to understand the most important scientific ideas. In the past, science education focused on memorization of concepts, facts, and procedures. Memorization is no longer enough; children need to acquire a deep and flexible understanding of the world of science. To this end, science should be taught so children are able to apply their knowledge in a wide variety of situations. Children often don't remember or understand teachers' explanations. Rather, children build their own knowledge as they link past learning to new experiences. For this reason, guiding students' beliefs toward scientifically accepted truths is often a long, challenging process.

    Science instruction needs to be substantial, based on how children learn, and centered on promoting children's understanding.

2. **Inquiry**. Children learn science best through inquiry. As they probe into scientific questions and problems of personal interest to them, children begin to acquire scientific knowledge and understand the process of inquiry. Both science content and the inquiry process are vital to a sound science education, and inquiry is the best way to help children learn both.

    The majority of class time should be spent in inquiry: children asking questions, designing and performing investigations, interpreting the results, and communicating their findings. The level of inquiry will depend on the age and abilities of the children.

3. **Hands-on**. Children begin to understand the natural world when they interact with it first-hand. While inquiry can, and should, include such second-hand sources as books, videos, computer programs, and experts, these need to be secondary to hands-on experimentation and observation.

    Inquiry should be primarily hands-on, with children planning, conducting, and reflecting on direct experiences.

FIGURE 10–4   Best practice in science education.

From: Glenview Public School District 34

4. **Choice**. When children are allowed choices, it results in greater motivation to learn, and a more complete understanding of science. Children often have limited interest in teacher-selected topics or activities. If given the choice of what to learn and how to learn it, however, most children become motivated to select topics and activities that make sense to them. Choice is the paramount way to differentiate instruction. It allows children to use different modes of learning and helps assure that each child is challenged. Choice does not mean free play, however. Teachers need to structure choice so important science topics are addressed in safe, practical ways.

   As much as practical, students should be allowed to choose science topics, formulate their own investigations, and draw their own conclusions.

5. **Reflection**. Reflection helps turn hands-on activities into concrete knowledge. Reflection is not "giving the answer" to hands-on activities, but rather an opportunity for each child to construct his own knowledge as he processes experiences and clarifies ideas. Reflection through discussion and writing also encourages effective communication about science.

   Reflection through class or small group discussion, learning logs, and science journals should be an integral and substantial part of each activity.

6. **Reading**. Reading of non-fiction reference and trade books is an important part of science education. Children need to be able to draw information from, and evaluate the validity of, outside sources. Books should not replace hands-on experiences, but can be used to initiate interest in a topic, extend knowledge gained hands-on, and provide answers to questions which don't lend themselves to first-hand experiences.

   Text and other second-hand resources should be subordinate to, and carefully intertwined with, first-hand experiences.

7. **Assessment**. Knowing where each child stands in relation to educational goals is essential for helping children reach those goals. Authentic assessments such as learning logs, inquiry reports, student observations, lab write-ups, and observation logs can all be used to assess progress. Assessment shouldn't be confined to concepts and facts but should include inquiry processes, reasoning ability, critical thinking, and attitudes toward science. In order to assess understanding rather than memorization, assessments must include non-routine applications of what children have learned.

   Assessment should be on-going and varied, and used to both guide and evaluate instruction.

FIGURE 10–4   Continued.

8. **Curriculum**. What teachers teach is as important as how they teach. Science curriculum should address the most important science concepts, including the inquiry process, life science, physical science, earth science, and technological design. It should be of an appropriate level of difficulty, so that it is understandable, challenging, and interesting for the students. Since children learn best through hands-on inquiry, topics should be able to be learned in this manner. The curriculum for each grade level needs to be carefully connected to, but not duplicated by, previous curriculum.

   Science curriculum must address important science. It should be both challenging, and within each child's potential for understanding.

9. **Teacher's role.** The teacher plays the preeminent role in her student's science education. Teachers do not have to have every science "answer." Usually "I don't know, how can we find out?" is preferable to providing answers to student questions. However, a firm knowledge of the scientific basics is needed for wise planning and teaching of science activities.

   Teachers need a solid foundation in the basics of science, deep knowledge of how children learn, and a positive, inquiring attitude toward science.

FIGURE 10–4   Continued.

what could be done differently next time. This continuous read/teach/discuss cycle can lead to long-lasting changes in how you and your colleagues teach as well as in your principal's view of math and science education.

The following are excellent books for book clubs.

• *Principles and Standards for School Mathematics* (National Council of Teachers of Mathematics 2000), *National Science Education Standards* (National Research Council 1996), *Benchmarks for Science Literacy* (Project 2061, American Association for the Advancement of Science 1993).

The national math and science education standards are authorative, informative, and easy to read. What's more, they are free; the References explain how to access them online. Much of *Principles and Standards for School Mathematics* is divided by age level, so there is a lot of specific information on math at the K–2 level. The math standards are organized around math topics, and they are rich in examples of ways to teach each topic. The *National Science Education Standards* include information on teaching, assessment, and science content. Again, specific classroom techniques are included, as well as broader principles. The science content section for younger children is written for the K–4 range. *Benchmarks for Scientific Literacy* focuses on science content, including what children should learn from kindergarten to second grade. Although it is less focused on teaching than the *National Science Education Standards*, it is valuable for highlighting the research on what children at each age level can and cannot understand; therefore, it is an essential resource for planning science curriculum.

- *Young Children Reinvent Arithmetic: Implications of Piaget's Theory* (Kamii 2000).

The second edition of this book provides a theoretical foundation for the teaching and learning of logic, number, and the operations. Aside from theory, however, it is an excellent source for math games, how to use word problems, and taking advantage of math situations outside of math class.

- *Best Practice: New Standards for Teaching and Learning in American Schools* (Zemelman, Daniels, and Hyde 1998).

In the second edition of this book, the authors outline what they view as the best educational practices. It is divided into each of the curriculum areas, including math and science. For each area it outlines and illustrates best practice; then it includes detailed snapshots of situations in real classrooms that exemplify best practice in action. There are also sections on how parents can help and the role of principals in promoting best practice. This authorative book cites a lot of research, but it is generally free from lingo; this makes it the best book among all cited here for parent book clubs also. It also includes extensive suggestions for further reading.

- *Developing Logical Thinking in Children* (Phillips 1991).

This book lays the groundwork for a developmentally appropriate math and science classroom for young children. It includes sections on managing a hands-on classroom, selecting materials, assessment (including excellent interviews), and nearly 400 pages of classroom activities. This book is essential for teachers interested in student-directed workshops.

- *Elementary Mathematics and Science Methods: Inquiry Teaching and Learning* (Foster 1999a).

This book is an excellent guide to constructivist math and science teaching for K–8 teachers. It is rich in classroom activities, including questions to ask as you conference with children. Quotes from the national standards helps to illustrate the standards' connections with constructivist teaching. This book also includes a section on assessing both learning and teaching.

Afterword

Encountering your own ideas outside of their usual environment can be startling. I experienced this phenomenon as I walked with my family in the Chicago Botanic Gardens on a sunny May afternoon. While my children played on a bridge that led over a lagoon, I overheard the conversation of a young girl and her father as they gazed into the water. He was telling her about the fish nests—circular patches that fish clear of vegetation as a place to lay eggs—that were visible on the lagoon floor. "Eggs in a nest?" the girl mused. "That's just like birds." Then, with the rush of joy that comes with sudden enlightenment: "Hey, I'm so smart!"

This is the kind of learning that math and science workshops can create: natural, and dynamic, built upon children's own interests, knowledge, and development. This kind of learning springs from children's curiosities as they immerse themselves in their own worlds. Knowledgeable, caring teachers play an essential role in this process, and like fiery Atlantic splashdowns, their impact is immeasurable.

References

Ames, C., and J. Archer. 1988. "Achievement goals in the classroom: Students' learning strategies and motivation processes." *Journal of Educational Psychology* 80: 260–267.

Andrews, A. G., and P. R. Trafton. 2002. *Little Kids—Powerful Problem Solvers: Math Stories from a Kindergarten Classroom.* Portsmouth, N.H.: Heinemann.

Atwell, N. 1987. *In the Middle: Writing, Reading, and Learning with Adolescents.* Portsmouth, N.H.: Boynton/Cook.

Ball, D. L. 1993. "Halves, pieces, and twoths: Constructing and using representational contexts in teaching fractions." In *Rational Numbers: An Integration of Research* edited by T. P. Carpenter, E. Fennema, and T. A. Romberg (Eds.) (157–195.) Hillsdale, N.J.: Erlbaum.

Baroody, A. J. 1998. *Fostering Children's Mathematical Power: An Investigative Approach to K–8 Mathematics Instruction.* Mahwah, N.J.: Lawrence Erlbaum Associates, Inc.

Baroody, A. J. 1999. Children's Relational Knowledge of Addition and Subtraction. *Cognition and Instruction* 17 (2): 175.

Battista, M. 1999. The Mathematical Miseducation of America's Youth: Ignoring Research and Scientific Study in Education. *Phi Delta Kappan* 80 (6): 424–433. Available online at http://www.pdkintl.org/kappan/kbat9902.htm.

Benware, C. A., and E. L. Deci. 1984. Quality of Learning with an Active Versus Passive Motivational Set. *American Educational Research Journal* 21: 755–765.

Bereiter. C., and M. Scardamalia. 1987. *The Psychology of Written Communication.* Hillsdale, N.J.: Lawrence Erlbaum Associates.

Berg, C. A., and D. G. Phillips. 1994. "An Investigation of the Relationship Between Logical Thinking Structures and the Ability to Construct and Interpret Line Graphs. *Journal of Research in Science Teaching* 3 (4): 323–344.

Black, P., and D. Wiliam, 1998a. "Assessment and Classroom Learning." *Assessment in Education* (March 1998): 7–74.

Black, P., and D. Wiliam. 1998b. Inside the Black Box: Raising Standards Through Classroom Assessment. *Phi Delta Kappan* 80 (2): Available online at <http://www.pdkintl.org/kappan/kbla9810.html>

Borasi, R., and B. J. Rose. 1989. Journal Writing and Mathematics Instruction. *Education Studies in Mathematics* 20: 347–65.

Boyd, S. E., and K. D. George, 1971. "The Effect of Science Inquiry on the Abstract Categorization Behavior of Deaf Children." Final report. ERIC document no. EC 041 400.

Bredderman, T. 1983. Effects of Activity-based Elementary Science on Student Outcomes: A Quantitative Analysis. *Review of Educational Research* 53 (4): 499–518.

Brownell, W. A. 1944. Rate Accuracy and Process in Learning. *Journal of Educational Psychology* 35: 321–337.

Brownell, W. A., and C. B. Chazal. 1935. The Effects of Premature Drill in Third-Grade Arithmetic. *Journal of Educational Research* 29: 17–28.

Butler, J. A. 1991. *Math Problem Solving Improvement: Troutdale Elementary School.* School Improvement Series, Snapshot #20. Northwest Regional Educational Laboratory. Available online at < http://www.nwrel.org/scpd/sirs/5/snap20.html>

Cai, J., J. Moyer, and N. Grochowski, 1997. Making the Mean Meaningful: Two Instructional Studies. Paper presented at the annual meeting of the American Educational Research Association, Chicago, Ill.

Cain-Caston, M. 1996. "Manipulative Queen." *Journal of Instructional Psychology* 23: 270–4.

Cajete, G. A. 1986. Science in a Native American Perspective (A Culturally Based Science Education Curriculum). Ph.D. dissertation, International College/William Lyon University, San Diego.

Calkins, L. M. 1986. *The Art of Teaching Writing.* Portsmouth, N.H.: Heinemann.

Carpenter, T. P., E. Fennema, M. L. Franke, L. Levi, and S. B. Empson, 1999. *Children's Mathematics: Cognitively Guided Instruction.* Portsmouth, N.H.: Heinemann.

Carpenter, T. P., L. F. Franke, V. R. Jacobs, E. Fennema, and S. B. Empson. 1998. A Longitudinal Study of Invention and Understanding in Children's Multidigit Addition and Subtraction. *Journal for Research in Mathematics Education* 29 (1): 3–20.

Carpenter, T. P., E. Fennema, P. L. Peterson, C. Chiang, and M. Loef. 1989. "Using Knowledge of Children's Mathematics Thinking in Classroom Teaching: An Experimental Study." *American Educational Research Journal* 26: 499–532.

Carr, J. F., and H. E. Harris. 2001. *Succeeding with Standards: Linking Curriculum, Assessment, and Action Planning.* Alexandria, Va.: Association for Supervision and Curriculum Development.

Carroll, W. M., and D. Porter. 1997. "Invented Strategies Can Develop Meaningful Mathematical Procedures." *Teaching Children Mathematics* 3 (7): 370–374.

Chang, C. Y., and S. L. Mao. 1999. "Comparisons of Taiwan Science Students' Outcomes with Inquiry-Group Versus Traditional Instruction." *The Journal of Educational Research*, 92 (6): 340–346.

Chinn, C. F., and W. F. Brewer. 1993. "The Role of Anomalous Data in Knowledge Acquisition: A Theoretical Framework and Implications for Science Instruction." *Review of Educational Research* 63 (1): 1–49.

Cobb, P., T. Wood, E. Yackel, and M. Perlwitz. 1992. "A Follow-up Assessment of a Second-Grade Problem-Centered Mathematics Project." *Educational Studies in Mathematics* 23 (5): 483–504.

Cobb, P., T. Wood, E. Yackel, J. Nicholls, G. Wheatley, B. Trigatti, and M. Perlwitz. 1991. "Assessment of a Problem-Centered Second Grade Mathematics Project." *Journal for Research in Mathematics Education* 22: 3–29.

Cohen, H. G. 1984. "The Effects of Two Teaching Strategies Utilizing Manipulatives on the Development of Logical Thought. *Journal of Research in Science Teaching* 21: 769–78.

Cohen, H. G. 1992. Two Teaching Strategies: Their Effectiveness with Students of Varying Cognitive Abilities. *School Science and Mathematics* 92 (3): 126–132.

Cohen, H. G. 1983. "A Comparison of the Effect of Two Types of Student Behavior with Manipulatives on the Development of Projective Spatial Structures." *Journal of Research in Science Teaching* 20 (9): 875–883.

Cronin-Jones, L. L. 2000. "The Effectiveness of Schoolyards as Sites for Elementary Science Instruction." *School Science and Mathematics* 100 (4): 203–211.

Dalton, B., C. C. Morocco, T. Tivnan, and P. L. R. Mead. 1997. Supported Inquiry Science: Teaching for Conceptual Change in Urban and Suburban Science Classrooms. *Journal of Learning Disabilities* 30 (6): 670–684.

Davison, D., and D. Pearce. 1990. *Perspectives on Writing Activities in the Mathematics Classroom.* ERIC document no. EJ 464 651.

Deci, E. L., and R. M. Ryan. 1985. *Intrinsic Motivation and Self-Determination in Human Behavior.* New York: Plenum Press.

Dossey, J. D., I. V. Mullis, M. M. Lindquist, and D. L. Chambers, 1988. *The Mathematics Report Card: Are We Measuring Up?* Princeton, N.J.: Educational Testing Service. ERIC document no. ED 300 206.

Everyday Learning Corporation. 1989. *Everyday Mathematics: The University of Chicago School Mathematics Project.* Chicago: Everyday Learning Corporation.

Everyday Learning Corporation. 2000. *Everyday Mathematics: Setting up an Everyday Mathematics Activity Night.* Available online at http://www.everydaylearning.com/em/infofor/hs/hs-mathnight.html

Foster, G. W. 1999a. *Elementary Mathematics and Science Methods: Inquiry Teaching and Learning.* Belmont, Calif.: Wadsworth Publishing Company.

Foster, G. W. 1999b. "Reflections on Teacher Philosophies and Teaching Strategies Upon Children's Cognitive Structure Development—Reflection I." A paper presented at the annual meeting of the Association for the Education of Teachers in Science. Austin, Texas, January 1999.

Foster, G. W., and A. Heiting. 1994. Embedded assessment. *Science and Children* 23 (2): 30–33.

Franke, M. L. and D. A. Carey. 1997. "Young Children's Perceptions of Mathematics: in Problem Solving Environments." *Journal for Research in Mathematics Education* 28, (1): 8–25.

Friel, S. N., F. R. Curcio, and G. W. Bright. 2001. "Making Sense of Graphs: Critical Factors Influencing Comprehension and Instructional Implications." *Journal of Research in Mathematics Education* 32 (2): 124–158.

Fuson, K. C., S. T. Smith, and A. M. L. Cicero. 1997. "Supporting Latino First Graders' Ten-Structured Thinking in Urban Classrooms." *Journal for Research in Mathematics Education 28* (6): 738–766.

Gallagher, S. A., W. J. Stiepen, and H. Rosenthal. 1992. "The Effects of Problem Based Learning on Problem Solving." *Gifted Child Quarterly* 36 (4): 195–200.

Gallagher, J. J. 2000. "Teaching for Understanding and Application of Science Knowledge." *School Science and Mathematics* 100 (6): 310–318.

Glasson, G. E., and R. V. Lakik. 1993. "Reinterpreting the Learning Cycle from a Social Constructivist Perspective. A Qualitative Study of Teachers' Beliefs and Practices." *Journal of Research in Science Teaching* 30: 187–207.

Glenview Public School District 34. 1996. *Curricular Expectations, K–3 Mathematics, 1996–1997.* Glenview, Ill.: Glenview Public School District 34.

Graves, D. H. 1983. *Writing: Teachers and Children at Work.* Portsmouth, N.H.: Heinemann.

Haury, D. L. 1993. "Teaching Science Through Inquiry." ERIC/CMSMEE Digest. ERIC document no. ED 359 048.

Henke, R. R., X. Chen, and G. Goldman. 1999. "What Happens in Classrooms? Instructional Practices in Elementary and Secondary Schools: 1994–1995." *Education Statistics Quarterly* 1 (2): 7–12.

Heuser, D. 1996. The Effects of Two Instructional Techniques (Object Exploration with Teacher Interaction, and Object Exploration Without Teacher Interaction) on First Grade Students' Growth Along the Pre-Classification – Class Inclusion Continuum. Master's Thesis, DePaul University, Chicago.

Heuser, D. 1997. "Toys in the Classroom: Developing Classification Skills Through Object Exploration." *Spectrum: The Journal of the Illinois Science Teachers Association* 24: 27–35.

Heuser, D. 1999. Reflections on Teacher Philosophies and Teaching Strategies upon Children's Cognitive Structure Development-Reflection II. A paper presented at the annual meeting of the Association for the Education of Teachers in Science. Austin Texas, January 1999. Available online at http://www.ed.psu.edu/CI/journals/1999aets/99file1.asp.

Heuser, D. 2000a. "Mathematics Workshop: Mathematics Class Becomes Learner Centered." *Teaching Children Mathematics* 6 (5): 288–295.

Heuser, D. 2000b. "Reworking the Workshop for Math and Science." *Educational Leadership* 58 (1): 34–37.

Heuser, D., and G. W. Foster. In preparation. A Research-Based Mathematics Program and Its Effects upon Conceptual Understanding and Cognitive Development.

Hiebert, J., and T. P. Carpenter. 1982. "Piagetian Tasks as Readiness Measures in Mathematics Instruction: A Critical Review." *Educational Studies in Mathematics* 13: 329–345.

Hiebert, J., T. P. Carpenter, and J. M. Moser. 1982. "Cognitive Development and Children's Solutions to Verbal Arithmetic Problems." *Journal of Research in Mathematics Education* 13 (2): 83–98.

Hiebert, J., and D. Wearne. 1993. Instructional Tasks, Classroom Discourse and Students' Learning in Second-Grade Arithmetic. *American Educational Research Journal* 30 (2): 393–425.

Hiebert, J., and D. Wearne, 1996. Instruction, Understanding, and Skill in Multidigit Addition and Subtraction. *Cognition and Instruction* 14: 251–283.

Hiebert, J., D. Wearne, and S. Taber. 1991. Fourth Graders' Gradual Construction of Decimal Fractions During Instruction in Different Physical Representations. *Elementary School Journal* 91 (4): 321–41.

Hoffer, T. B., and A. Gamoran. 1993. *Effects of Instructional Differences Among Ability Groups on Student Achievement in Middle-School Science and Mathematics.* ERIC no. SE 053 810.

Illinois State Board of Education. 1997. *Illinois Learning Standards (1997).* Springfield, Ill: Illinois State Board of Education.

International Association for the Evaluation of Educational Achievement. 1995. *Data Almanacs for Achievement Items: Distribution of Responses for TIMSS Items. Fourth Grade Science.* Available online at http://www.isc.bc.edu/timss1995i/data_almanacs_95.html.

International Association for the Evaluation of Educational Achievement. 1998.*TIMSS Mathematics Items: Released Set for Population 1 (Third and Fourth Grade).* Available online at http://www.timss.org/TIMSS1/items.html.

Issacs, A. C., and W. M. Carroll. 1999. "Strategies for Basic-Facts Instruction." *Teaching Children Mathematics* 5 (9): 508–515.

Jurdak, M., and R. Abu Zein. 1998. "The Effect of Journal Writing on Achievement In and Attitudes Toward Mathematics." *School Science and Mathematics* 98 (8): 412–419.

Kahle, J. B., and A. Damnjanovic. 1994. "The Effect of Inquiry Activities on Elementary Students' Enjoyment, Ease, and Confidence in Doing Science: An Analysis by Sex and Race." *Journal of Women and Minorities in Science and Engineering* 1 (1): 17–28.

Kamii, C., with L. L. Joseph. 1989. *Young Children Continue to Reinvent Arithmetic—2nd Grade: Implication of Piaget's Theory.* New York: Teachers College Press.

Kamii, C., with L. B. Housman. 2000. *Young Children Reinvent Arithmetic: Implications of Piaget's Theory* (2nd ed.). New York: Teachers College Press.

Kamii, C., and A. Dominick. 1998. "The Harmful Effects of Algorithms in Grades 1–4." In *The Teaching and Learning of Algorithms in School Mathematics*, edited by L. J. Morrow, and M. J. Kenney. *1998 Yearbook*. Reston, Va. National Council of Teachers in Mathematics.

Keys, C. W. 1999. Revitalizing instruction in scientific genres: Connecting knowledge production with writing to learn in science. *Science Education, 83*, 115–130.

Leap, W. J. 1982. *Dimensions of Math Avoidance Among American Indian Elementary School Students*. Washington, D.C.: National Institute of Education . ERIC no. ED 244 748.

Lee, O., and S. Paik, 2000. "Conceptions of Science Achievement in Major Reform Documents." *School Science and Mathematics* 100 (1): 16–26.

Lindquist, M. M., and V. L. Kouba. 1989. "Measurement." In *Results from the Fourth Mathematics Assessment of the National Assessment of Education Progress*, edited M. M. Lindquist. (35–43). Reston, Va.: National Council of Teachers of Mathematics.

Lockwood, A. 1992a. "The De Facto Curriculum." *Focus in Change* 6, 8–11.

Lockwood, A. 1992b. "Whose Knowledge Do We Teach?" *Focus in Change* 6: 3–7.

Marzano, R. J. 2000. *Transforming Classroom Grading*. Alexandria, Va.: Association for Supervision and Curriculum Development.

Mather, J. R. C. 1997. "How Do American Indian Fifth and Sixth Graders Perceive Mathematics and the Mathematics Classroom?" *Journal of American Indian Education* 36: 9–18.

Mattheis, F. E., and G. Nakayama. 1988. *Effects of a Laboratory-Centered Inquiry Program on Laboratory Skills, Science Process Skills and Understanding of Science Knowledge in Middle Grades Students*. ERIC document no. SE 050 605.

McClain, K., P. Cobb, and J. Bowers. 1998. "A Contextual Investigation of Three-Digit Addition and Subtraction." In *The Teaching and Learning of Algorithms in School Mathematics. 1998 Yearbook*, edited by L. J. Morrow, and M. J. Kenney. Reston, Va.: National Council of Teachers in Mathematics.

McDavitt, D. S. 1994. *Teaching for Understanding: Attaining Higher Order Learning and Increased Achievement Through Experiential Instruction*. ERIC document no. ED 374 093.

Meece, J. L. 1991. "The Classroom Context and Students' Motivational Goals." In *Advances in Motivation and Achievement: A Research Annual* (Vol. 7, pp. 261–85), edited by M. L. Maehr and P. R. Pintrich. Greenwich, Conn.: JAI Press.

Meyer, D. K., J. C. Turner, and C. A. Spencer. 1997. "Challenge in a Mathematics Classroom: Students' Motivation and Strategies in Project-Based Learning." *Elementary School Journal, 97 (5):* 501–21.

Mitchell, J. H., E. F. Hawkins, F. B. Stancavage, and J. A. Dossey, 2000. Estimation Skills, Mathematics-in-Context, and Advanced Skills in Mathematics." *Education Statistics Quarterly* 2 (1): 29–33.

Mitchell, J. H., E. F. Hawkins, P. M. Jakwerth, F. B. Stancavage, and J. A. Dossey, 1999. "Student Work and Teacher Practices in Mathematics." *Educational Statistics Quarterly* 1 (2): 39–43.

Moscovici, H., and Nelson, T. H. 1998. "Shifting from Activitymania to Inquiry." *Science and Children* 40, 14–17.

Narrode, R., J. Board, and L. Davenport. 1993. "Algorithms Supplant Understanding: Case Studies of Primary Students' Strategies for Double-Digit Addition and Subtraction." In *Proceedings of the Fifteenth Annual Meeting, North American Chapter of the International Group for the Psychology of Mathematics Education*, Volume 1, edited by J. R. Becker and B. J. Pence (pp. 254–60). San Jose, Calif.: San Jose State University, Center for Mathematics and Computer Science Education.

National Council of Teachers of Mathematics. 2000. *Principles and Standards for School Mathematics.* Reston, Va.: National Council of Teachers of Mathematics. Available online at http://www.standards.nctm.org/index.htm.

National Research Council. 1989. *Everybody Counts: A Report to the Nation on the Future of Mathematics Education.* Washington, D.C.: National Academy Press.

National Research Council. 1996. *National Science Education Standards.* Washington, D.C.: National Academy Press. Available online at http://www.nap.edu/readingroom/books/nses/.

National Research Council. 2000. *Inquiry and the National Science Education Standards: A Guide for Teaching and Learning.* Edited by S. Olson, and S. Loucks-Horsley. Committee on the Development of an Addendum to the National Science Education Standards on Scientific Inquiry, National Research Council. Washington, D.C.: National Academy Press. Available online at http://www.bob.nap.edu/html/inquiry_addendum/notice.html.

National Research Council. 2001. *Classroom Assessment and the National Science Education Standards.* Edited by J. M. Atkin, P. Black, and J. Coffey. Committee on Classroom Assessment and the *National Science Education Standards*, Center for Education, National Research Council. Washington, DC: National Academy Press. Available online at http://www.books.nap.edu/html/classroom_assessment/.

National Science Board. 1991. *Science and Engineering Indicators—1991.* Washington, D.C.: U.S. Government Printing Office (NSB 91-1).

National Science Foundation. 2000. *Foundations Volume 1: A Monograph for Professionals in Science, Mathematics, and Technology Education. The Challenge and Promise of K–8 Science Reform.* Division of Elementary, Secondary, and Informal Education, Directorate for Education and Human Resources, National Science Foundation. Washington, D.C.: U.S. Government Printing Office (NSF-9776). Available online at http://www.nsf.gov/subsys/ods/getpub.cfm?ods_key=nsf9776

O'Brien, T. C. 1999. "Parrot Math." *Phi Delta Kappan* 80 (6): 434–438. Available online at http://www.pdkintl.org/kappan/kobr9902.htm.

O'Sullivan, C. Y., and A. R. Weiss. 1999. Student Work and Teacher Practices in Science. *Education Statistics Quarterly* 1 (2): 44–45.

Palincsar, A. S., C. Anderson, and Y. M. David, 1993. "Pursuing Scientific Literacy in the Middle Grades Though Collaborative Problem Solving." *The Elementary School Journal* 93: 643–658.

Peasley, K. L., C. L. Rosean, and K. J. Roth. 1992. *Why Did We Do All This Writing and Talking If You Already Knew the Answer? The Role of the Learning Community in Constructing Understanding in an Elementary Science Class.* Elementary Subjects Center, Series No. 61. East Lansing, Mich: Center for the Learning and Teaching of Elementary Subjects, Institute for Research on Teaching.

Phillips, D. R. 1991. *Developing Logical Thinking in Children.* Coralville, Ia.: Insights.

Phillips, D. R., D. G. Phillips, with G. Melton and P. Moore. 1994. "Beans, Blocks and Buttons: Developing Thinking." *Educational Leadership* 51 (5): 50–53.

Phillips, D. G., and D. R. Phillips, 1996. *Structures of Thinking: Concrete Operations.* (2nd ed.). Dubuque, Ia.: Kendall/Hunt.

Phillips, D. G. 1992. *Sciencing: Toward Logical Thinking.* North Liberty, Ia.: Insights.

Piaget, J. 1964. "Cognitive Development in Children." *Journal of Research in Science Teaching* 2: 176–186.

Popham, W. J. 1999. Why Standardized Tests Don't Measure Educational Quality. *Educational Leadership* 56 (6): 8–15.

Project 2061, American Association for the Advancement of Science. 1993. *Bench-marks for Science Literacy*. New York: Oxford University Press, Inc. Available on-line at http://www.project2061.org/tools/benchol/bolframe.htm.

Rakow, S. J. 1986. *Teaching Science as Inquiry*. Fastback 246. Bloomington, Ind.: Phi Delta Kappa Educational Foundation. ERIC document no. ED 275 506.

Rathmell, E. C. 1978. "Using Thinking Strategies to Learn Basic Facts." In *Developing Computational Skills. The 1978 Yearbook of the National Council of Teachers of Mathematics,* edited by Marilyn Suydam. Reston, Va.: National Council of Teachers of Mathematics.

Resnick, L. B., V. L. Bill, S. B. Lesgold, and M. N. Leer. 1991. "Thinking in Arithmetic Class." In *Teaching Advanced Skills to At-risk Students: Views from Research and Practice* (p. 27–67), edited by B. Means, C. Chelemer, and M. S. Knapp. San Francisco: Jossey-Bass.

Rivard, L. P., and S. B. Straw, (2000). "The Effect of Talk and Writing on Learning Science: An Exploratory Study." *Science Education* 84: 566–593.

Romberg, T. A. 2000. "Changing the Teaching and Learning of Mathematics." *Australian Mathematics Teacher* 56 (4): 6–9.

Rosebery, A. S., B. Warren, and F. R. Conant. 1992. Appropriating Scientific Discourse: Findings from Language Minority Classrooms. *The Journal of Learning Sciences* 2 (1): 61–94.

Ryan, A. M., L. Hicks, and C. Midgely, 1997. "Social Goals, Academic Goals, and Avoiding Seeking Help in the Classroom." *Journal of Early Adolescence* 17 (2): 152–171.

Schauble, L., L. E. Klopfer, and K. Raghavan, 1991. "Students' Transition from an Engineering Model to a Science Model of Experimentation." *Journal of Research in Science Teaching* 28: 859–882.

Schmidt, W. H., C. C. McKnight, and A. A. Raizen, 1997. *Splintered Vision: An Investigation of U.S. Mathematics and Science Education.* Norwell, Mass.: Kluwer Academic.

Scruggs, T. E., M. A. Mastropieri, J. P. Bakken, and F. J. Brigham. 1993. Reading Versus Doing: The Relative Effects of Textbook-Based and Inquiry-Oriented Approaches to Science Learning in Special Education Classrooms. *The Journal of Special Education* 27: 1–15.

Shaw, E. L., and M. M. Hatfield. 1996. A Survey of the Use of Science Manipulatives in Elementary Schools. Paper presented at the annual meeting of the Mid-South Education Research Association (Tuscalloosa, Ala., November 6, 1996).

Shaw, K. L., and E. Jakubowski. 1991. Teachers Changing for Changing Times. *Focus on Learning Problems in Mathematics* 13 (4): 13–20.

Shymansky, J. A., L. V. Hedges, and G. Woodworth. 1990. A Reassessment of the Effects of 60's Science Curricula on Student Performance. *Journal of Research in Science Teaching* 27 (2): 127–144.

Silver, J. W. 1999. A Survey on the Use of Writing-to-Learn in Mathematics Classes. *Mathematics Teacher* 92 (5): 388–389.

Silverstein, S. 1987. *The Giving Tree*. New York: HarperCollins Children's Books.

Sowell, E. J. 1989. "Effects of Manipulative Materials in Mathematics Instruction." *Journal for Research in Mathematics Education* 20 (5): 498–512.

Stevenson, H. W. 1998. "A TIMSS Primer. Lessons and Implications for U.S. Education." *Fordham Report* 2 (7): Available online at <http://www.edexcellence.net/library/timss.html>

Stigler, J. W., and J. Hiebert, 1997. Understanding and Improving Classroom Mathematiacs Instruction: An Overview of the TIMSS Video Study. *Phi Delta Kappa* 79 (1):14–21.

Stigler, J. W., and M. Perry. 1990. "Mathematics Learning in Japanese, Chinese, and American Classrooms." In *Children's Mathematics. New Directions for Child Development, No. 41,* (pp. 27–54), edited by G. B. Saxe and M. Gearhart. San Francisco: Jossey-Bass.

Stipek, D. 1996. "Motivation and instruction." In *Handbook of Educational Psychology* (85–113), edited by D. C. Berliner and R. C. Calfee. New York: Macmillan.

Stipek, D., J. M. Salmon, K. B. Givvin, E. Kazemi, G. M. Saxe, and V. L. MacGyvers. 1998. The Value (and Convergence) of Practices Suggested by Motivation Research and Promoted by Mathematics Education Reformers. *Journal for Research in Mathematics Education* 29 (4): 465–488.

Stohr-Hunt, P. M. 1996. "An Analysis of Frequency of Hands-on Experience and Science Achievement." *Journal of Research in Science Teaching* 33 (1): 101–109.

Sunflower, C., and L. W. Crawford. 1985. *How Frequently Are Elementary Students Writing?* ERIC document no. ED 272 895.

Suydam, M. N. 1985. *Research on Instructional Material for Mathematics.* ERIC SMEAS Special Digest No. 3, 1985. ERIC document no. ED 276 569.

Suydam, M. N., and J. L. Higgins. 1977. *Activity-Based Learning in Elementary School Mathematics: Recommendations from Research.* ERIC document no. ED 144 840.

U.S. Department of Education, National Center for Education Statistics. 1996. *Pursuing Excellence.* Washington, D.C.: U.S. Government Printing Office. Available online at http://www.nces.ed.gov/timss/.

U.S. Department of Education, National Commission on Mathematics and Science Teaching for the 21st Century. 2000. *Before It's Too Late.* Washington, D.C.: U.S. Government Printing Office. Available online at http://www.ed.gov/americacounts/glenn/.

Van Tassel-Baska, J. 1998. *Planning Science Programs for High Ability Learners.* ERIC Digest E546. Reston, Va.: ERIC Clearinghouse on Disabilities and Gifted Education.

Van Tassel-Baska, J., G. Bass, R. Ries, D. Poland, and L. Avery. 1998. A National Study of Science Curriculum Effectiveness with High Ability Students. *Gifted Child Quarterly* 42 (4): 200–211.

Wafler, E. S. 2001. Inspired Inquiry: Using Candy and Containers of Water to Spark Students' Interest in Investigating. *Science and Children* (January): 28–31.

Wason-Ellam, L. 1987. Writing as a Tool for Learning: Math Journals in Grade One. Paper presented at the Sixth Annual Meeting of the National Council of Teachers of English (Louisville, KY, March 26–28, 1987).

Wheatley, G. H. 1992 "The Role of Reflection in Mathematics Learning." *Educational Studies in Mathematics* 23 (5): 529–541.

Wilson, L. D., and R. K. Blank. 1999. *Improving Mathematics Education Using Results NAEP and TIMSS.* ERIC document no. ED 431 001.

Wise, K. C. 1996. "Strategies for Teaching Science: What Works?" *The Clearing House* 69: 337–338.

Yang, M. T. L., and P. Cobb. 1995. A Cross-Cultural Investigation into the Development of Place-Value Concepts of Children in Taiwan and the United States. *Educational Studies in Mathematics* 28 (1): 1–33.

Zemelman, S., H. Daniels, and A. Hyde. 1998. *Best Practice. New Standards for Teaching and Learning in American Schools.* (2nd. Ed.). Portsmouth, N.H.: Heinemann.